200 Challenging Math Problems

Every 4th Grader Should Know

This book belongs to:

marcus

Grade: 5th

200 Challenging Math Problems

every 4th grader should know

New edition, 2017
Copyright Learn 2 Think Pte. Ltd.

Published by:
Learn 2 Think Pte. Ltd.

ISBN: 978-981-072-765-9

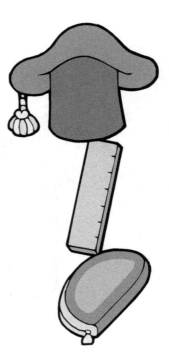

Master Grade 4 Math Problems

Introduction:

Solving math problems is core to understanding math concepts. When Math problems are presented as real-life problems students get a chance to apply their Math knowledge and concepts they have learnt. Word problems progressively develop a student's ability to visualize and logically interpret Mathematical situations.

This book provides numerous opportunities to students to practice their math skills and develop their confidence of being a lifelong problem solver. The multi-step problem solving exercises in the book involve several math concepts. Student will learn more from these problems solving exercises than doing ten worksheets on the same math concepts. The book is divided into 7 chapters. The last chapter of the book explains step wise solutions to all the problems to reinforce learning and better understanding.

How to use the book:

Here is a suggested plan that will help you to crack every problem in this book and outside.

Follow these 4 steps and all the Math problems will be a NO PROBLEM!

Read the problem carefully:

- ✎ What do I need to find out?
- ✎ What math operation is needed to solve the problem? For example addition, subtraction, multiplication, division etc.
- ✎ What clues and information do I have?
- ✎ What are the key words like sum, difference, product, perimeter, area, etc.?
- ✎ Which is the non-essential information?

Decide a plan

- ✎ Develop a plan based on the information that you have to solve the problem. Consider various strategies of problem solving:
- ✎ Drawing a model or picture
- ✎ Making a list
- ✎ Looking for pattern
- ✎ Working backwards
- ✎ Guessing and checking
- ✎ Using logical reasoning

Solve the problem:

Carry out the plan using the Math operation or formula you choose to find the answer.

Check your answer

- ✎ Check if the answer looks reasonable
- ✎ Work the problem again with the answer
- ✎ Remember the units of measure with the answer such as feet, inches, meter etc.

Master Grade 4 Math Problems

Note to the Teachers and Parents:

✐ Help students become great problem solvers by modelling a systematic approach to solve problems. Display the 'Four step plan of problem solving' for students to refer to while working independently or in groups.

✐ Emphasise on some key points:

✐ Enable students to enjoy the process of problem solving rather than being too focused on finding the answers.

✐ Provide opportunities to the students to think; explain and interpret the problem.

✐ Lead the student or the group to come up with the right strategy to solve the problem.

✐ Discuss the importance of showing steps of their work and checking their answers.

✐ Explore more than one possible solution to the problems.

✐ Give a chance to the students to present their work.

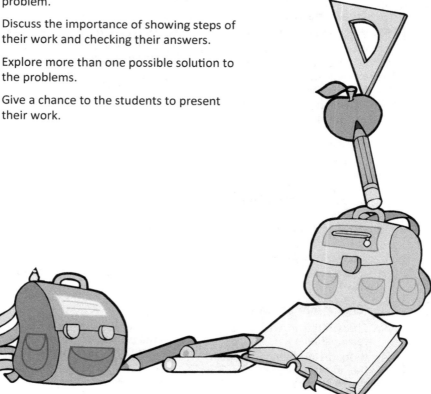

Master Grade 4 Math Problems

Introduction:

Solving math problems is core to understanding math concepts. When Math problems are presented as real-life problems students get a chance to apply their Math knowledge and concepts they have learnt. Word problems progressively develop a student's ability to visualize and logically interpret Mathematical situations.

This book provides numerous opportunities to students to practice their math skills and develop their confidence of being a lifelong problem solver. The multi-step problem solving exercises in the book involve several math concepts. Student will learn more from these problems solving exercises than doing ten worksheets on the same math concepts. The book is divided into 7 chapters. The last chapter of the book explains step wise solutions to all the problems to reinforce learning and better understanding.

How to use the book:

Here is a suggested plan that will help you to crack every problem in this book and outside.

Follow these 4 steps and all the Math problems will be a NO PROBLEM!

Read the problem carefully:

- ✎ What do I need to find out?
- ✎ What math operation is needed to solve the problem? For example addition, subtraction, multiplication, division etc.
- ✎ What clues and information do I have?
- ✎ What are the key words like sum, difference, product, perimeter, area, etc.?
- ✎ Which is the non-essential information?

Decide a plan

- ✎ Develop a plan based on the information that you have to solve the problem. Consider various strategies of problem solving:
- ✎ Drawing a model or picture
- ✎ Making a list
- ✎ Looking for pattern
- ✎ Working backwards
- ✎ Guessing and checking
- ✎ Using logical reasoning

Solve the problem:

Carry out the plan using the Math operation or formula you choose to find the answer.

Check your answer

- ✎ Check if the answer looks reasonable
- ✎ Work the problem again with the answer
- ✎ Remember the units of measure with the answer such as feet, inches, meter etc.

Master Grade 4 Math Problems

Note to the Teachers and Parents:

- Help students become great problem solvers by modelling a systematic approach to solve problems. Display the 'Four step plan of problem solving' for students to refer to while working independently or in groups.

- Emphasise on some key points:

- Enable students to enjoy the process of problem solving rather than being too focused on finding the answers.

- Provide opportunities to the students to think; explain and interpret the problem.

- Lead the student or the group to come up with the right strategy to solve the problem.

- Discuss the importance of showing steps of their work and checking their answers.

- Explore more than one possible solution to the problems.

- Give a chance to the students to present their work.

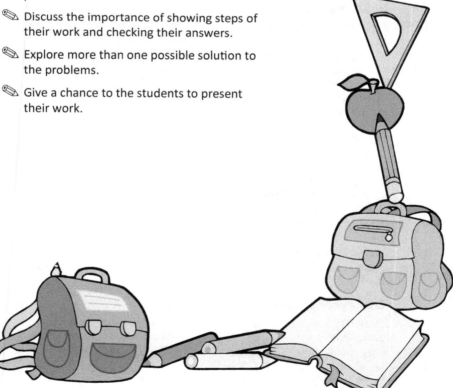

CONTENTS

Topics	**Page number**

I am a number less than 100. My units digit is a 4. The sum of my digits is an odd number. My tens digit is a multiple of 3. What number am I?

$$3 \times 4 = 12$$
$$4 \times 3 = 12$$
$$100 - 12 = 98$$
$$12 + 100 = 112$$

Answer: ....98.................

PROBLEM 2

I am a two digit number. If you round
me to the nearest hundred, my value
will increase by ten. What number am I?

Answer: 90

Add +, -, x, ÷ to make the equations work.
There can be more than 1 solution.

3 3 3 3 3 = 0
3 3 3 3 3 = 1
3 3 3 3 3 = 2
3 3 3 3 3 = 3
3 3 3 3 3 = 5
3 3 3 3 3 = 6
3 3 3 3 3 = 7
3 3 3 3 3 = 9

Answer: ....

9

PROBLEM 4

The odometer of my car indicates 187569.
All the digits of this number are different.
After how many more kilometers will this
happen again?

Answer: ....

I am an even number between 5 and 20. If you keep subtracting 5 from me, you will be left with 0. What number am I?

Answer: 10 and 20

Gavin set a secret security number to lock his phone. It is greater than 275 but less than 325; if you count by 5s you say its name; it can be divided evenly by 3 and 9. What is the secret number?

Answer: ....

Using the numbers 2, 3, and 6, what is the smallest number you can get if you multiply two of the numbers and add the third to the product? What is the greatest number you can get?

Answer: ....

PROBLEM 8

The number of coins in my wallet is greater than 30 but less than 50. When I arrange them in the groups of two, no coins are left behind. When I arrange them in the groups of 5, no coins are left behind either. How many coins do I have in my wallet?

Answer: ....

PROBLEM 9

I am a secret two digit odd number. I am less than 20 and the sum of my two digits is 10. What number am I?

Answer: ....

15

PROBLEM 10

Find a pair of numbers with:

a) a sum of 11 and a product of 24.

b) a sum of 40 and a product of 400.

c) a sum of 15 and a product of 54.

Answer: ....

I am a number between 190 and 207.
When divided by 4, the remainder is 2.
When divided by 5, the remainder is 1.
What number am I?

Answer: ....

PROBLEM 12

Ben counted up all the pets at a pet show and got a number. The number is greater than 195 but less than 300. It is a multiple of 10 and it can be divided evenly by 3. It has 9 as a factor. How many pets did Ben count at the pet show?

Answer: ....

PROBLEM 13

Given below is the area of some countries in square kilometers . Answer the following questions based on the information provided.

USA: 9,628,091
Russia: 17,097,242
China: 9,599,094
UK: 242,300
India: 3,287,293

Which of these countries has the smallest area?

Which of these countries has the largest area?

What is the difference between the areas of Russia and China?

Find the total area of all countries listed above?

Order these countries from the largest to the smallest areas?

Answer: ....

19

PROBLEM 14

Lisa, Amy, and Pete decided to race up the stairs that have 24 steps. Lisa takes 2 steps every second, Amy takes 3 steps every second while Pete takes 4 steps every second.

If Pete starts at the bottom, what step should Lisa and Amy start on so that all three finish in a tie?

Answer: ....

20

PROBLEM 15

There are 25 children standing in a straight line in Mrs. Smith's class. Every 4th child is wearing spectacles. Every 3rd child is a girl. Every 2nd child is wearing a white shirt. What can you say about the 12th, 18th, and the 21st child in Mrs. Smith's class?

Answer: ....

PROBLEM 16

According to the clock in Sam's room, it is 9:00 am on June 12th. If Sam turns the minute hand counter clockwise 30 rounds, what date and time will the clock show?

Answer: ....

PROBLEM 17

The following diagram shows the first three patterns of squares in a sequence. How many squares are there in the 50th pattern?

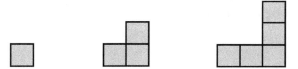

Answer: ....

PROBLEM 18

Katie placed 8 cubes together in a row on a table and sprayed paint on them. Every cube has 6 faces. The faces that touch the table and the ones that touch the next cube don't get painted. How many faces of the cubes get painted?

Answer: ....

PROBLEM 19

A street is 350 meters long. Palm trees are planted on both sides of the street from the beginning to the end of the street at 5 meters apart. How many palm trees are planted?

Answer: ....

PROBLEM **20**

Aaron has between 100 and 110 stickers in his collection. If he counts his stickers 2 at a time, he has 1 sticker left. If he counts his stickers 5 at a time, he has 2 stickers left . How many stickers are there in Aaron's collection?

Answer: ………………………

Claire thought of a number. She subtracted 99 from the number and added 57 to it. The final answer is 556. What number did Claire think of?

Answer: ....

PROBLEM 22

Judy had 2 chicken at her farm during the first week. At the end of the second week, she had 4 chicken. At the end of the third week, she had 8. If this pattern continues, how many chicken will be there at her farm by the end of the fifth week?

Answer: ....

Neil is making a house of cards by arranging them in the shape of a triangle. He places 1 card on the top, 4 cards in the row below, 7 cards in row 3 and so on. If the house of cards contains 6 rows, how many cards are there in the bottom row?

Answer: ....

PROBLEM 24

If you cut a piece of ribbon in the middle, you will have two pieces. If you put these two pieces together and cut them in the middle again, you will have 4 pieces. If you put these 4 pieces together and cut again in the middle, you will have 8 pieces and so on.

If you keep cutting in the same pattern, how many pieces will you have after 10 cuts?

Answer: ....

Russell had two dozen orange cookies. He ate four cookies on Sunday. On every day that followed, he ate the number of cookies that was two greater than the day before. On which day did Russell find that he did not have enough cookies to continue eating in the same pattern?

Answer: ....

Judy and Jane are playing a board game with four dice. Each player rolls four dice and uses the four numerals rolled to make 4-digit numbers. A numeral can only be used once in a number. Jane rolls 3, 5, and 6 and 9. Write all the possible numbers Jane can make using the four numerals. Which is the largest 4 digit number Jane can make?

Answer: ....

PROBLEM 27

In a marathon, Mike finished ahead of Fred. Emma finished after Fred. John finished after Agnes and Agnes finished after Emma. In what sequence did each of them finish the marathon?

Answer: ....

PROBLEM 28

Tim goes to the church every third day, Jason goes to the church every 4th day and Jacob goes every 6th day. If they all went to the church today, when will they all go to the church on the same day again?

Answer: ………………………

34

A packet of plastic forks contains 10 forks. A packet of plastic knives contains 12 knives. What is the fewest number of packets you would have to buy to have exactly the same number of forks as knives?

Answer: ....

PROBLEM 30

Nicole has between 70 to 100 stamps in her collection. When she divides them into groups of 2 or 3 or 7, there is always 1 stamp left. How many stamps does Nicole have in her collection?

Answer: ....

PROBLEM 31

There are 7 people in a meeting. If each person shakes hands with the other exactly once, how many handshakes are done?

Answer: ....

PROBLEM **32**

Mr David is building a fence around his field. He placed 17 fence posts 6 feet apart. What is the distance from the first fence post to the last?

Answer: ....

At the end of a game, Mary was 40 points ahead of Sandra. Ben was 35 points behind Jenny. Sandra was 10 points ahead of Jenny. Who won the game? In what order did the other players finish?

Answer: ....

PROBLEM 34

Helen had to arrange some chairs in a fixed number of rows. She estimated that there were more than 40 but less than 80 chairs. If she put 8 chairs in one row, she would have 1 chair left. If she put 7 chairs in one row, she would have 2 chairs left. How many chairs did Helen have to arrange?

Answer: ....

PROBLEM 35

Fill the blanks with +, -, x and ÷ signs to make the following equations work.

a) 1 __ 1 __ 1 __ 1 = 1

b) 5 __ 5 __ 5 __ 5 = 25

Answer: ....

41

PROBLEM 36

Sam, Carla and Sarah spent one afternoon collecting sea shells. Sam collected 11 sea shells. If we add the number of sea shells collected by Sam and Carla, the total would be 24. If we add the number of sea shells collected by Carla and Sarah, the total would be 25 shells. How many shells did each one of them collect?

Answer: ....

Lucy was 130 cm tall when she was 8 years old. In the next year she grew 5 cm, and the following year 3 cm less than the previous year. How tall was Lucy at the age of 10?

Answer: ....

PROBLEM 38

For a school picnic, Mrs Smith divided her students into 5 groups. In each group, there were 18 boys and 8 fewer girls than boys. How many students went for the school picnic?

Answer: ....

PROBLEM 39

Bill, Tim, and Alice went fishing. Bill caught 28 more fish than Alice and Alice caught 15 fish less than Tim. Tim caught 66 fish. How many fish did each one of them catch?

Answer: ....

PROBLEM 40

In a car park, there are 198 buses. There are 54 more buses than vans and 36 more motorcycles than vans. How many vehicles are there altogether?

Answer: ....

Mr James bought 19 crates of fruits. There were 64 fruits in each crate. If there were 320 apples, 414 pears and the remaining were mangoes; how many mangoes were there in all?

Answer: ....

PROBLEM 42

Sam, Hansel, and Amie had a collection of cards. Sam had 260 cards. Amie had twice as many cards as Sam. Hansel had as many cards as Sam and Amie together. How many cards did Hansel have?

Answer: …………………………

Mark grew some plants in his garden. He planted 6 rows of plants. The first row had 10 plants. The second row had twice as many plants as the first. The third and fifth rows had the sum of rows 1 and 2. The fourth and sixth rows had the sum of rows 1, 2, and 3. How many plants did Mark plant in his garden?

Answer: ....

Ryan bought 20 boxes of pencils. There were 24 pencils in each box. He gave 6 pencils to each of the 30 students in his class. How many pencils were left with Ryan?

Answer: ....

PROBLEM 45

Sam, Lucy, Emily and Sarah decided to go for a roller coaster ride. Emily went twice as many times as Lucy. Sam went four more times than Sarah but three less times than Emily. Sarah went 5 times altogether. How many times did Lucy go?

Answer: ....

PROBLEM 46

If 18 people share a barrel of apples equally, each person gets 12 apples. If there were 6 fewer people, how many apples would each person get?

Answer: ………………………….

PROBLEM 47

A machine can print 70 cards in 6 minutes. How many cards can it print in 2 hours?

Answer: ....

PROBLEM 48

Sam ordered 145 cans of coke on Monday. On Tuesday, he ordered 6 times as many cans of coke as he did on Monday. How many more cans of coke did Sam order on Tuesday than on Monday?

Answer: ....

PROBLEM 49

Amy bought 95 boxes of mangoes. There were 50 mangoes in each box. She gave 25 mangoes to her friends and repacked the rest into smaller boxes of 5 mangoes each.

a) How many small boxes of mangoes did Amy pack?

b) How much money did she earn if she sold each box for $15?

Answer: ....

Bernard has sixteen pets. All of his pets are cats and dogs. Bernard has four more cats than dogs. How many dogs does he have?

Answer: ....

PROBLEM 51

There are 6 large boxes and inside each large box there are 4 small boxes. Inside each of these small boxes there are 3 smaller boxes. Counting all sizes, how many boxes are there altogether?

Answer: ....

PROBLEM 52

Sarah, Nicole, Amanda, and Suzy have some Barbie dolls. Sarah has thrice as many dolls as Suzy. Suzy has eight more dolls than Nicole. Suzy has twenty eight dolls.
Amanda has four dolls less than Nicole. How many dolls does each girl have?

Answer: ....

PROBLEM 53

When wheel B rotates two times, wheel A rotates 6 times. When wheel A rotates 40 times, how many times would wheel B rotate?

Answer: ....

A box can hold 450 marbles. A container can hold 5 times as many marbles as a box. How many marbles can 5 boxes and 2 containers hold?

Answer: ....

PROBLEM 55

A buffet lunch costs $23 for an adult and $7 less for a child. Find the total amount of money that a family of 2 adults and 3 children would have to pay for the lunch.

Answer: ....

PROBLEM 56

Doris bought 8 packets of tomatoes and gave them to 25 students. When she gave 4 tomatoes to each of them, she had 4 tomatoes left. How many tomatoes were there in each packet?

Answer: ....

PROBLEM 57

A shopkeeper sold 230 potatoes on Monday and twice as many potatoes on Tuesday and had 150 potatoes left. If he wanted to sell equal number of potatoes within 2 days, how many potatoes should he sell on each day?

Answer: ....

9 packets of sweets cost $72 and 3 packets of biscuits cost $12. Alice wanted to buy 4 packets of sweets and 2 packets of biscuits. Find the total amount of money that she had to pay for the items.

Answer: ....

Florence bought 18 packets of apples weighing 5 kg each. She then packed them into plastic bags of 2 kg each. How many plastic bags did she use?

Answer: ....

PROBLEM 60

8 people paid $90 each for a dinner party. The party organizers were still short of $24. What was the actual amount of money that each person should have paid to cover the cost of the entire dinner party?

Answer: ....

PROBLEM 61

Tracy bought 2 similar watches and had $150 left. She spent 3 times the amount of money left with her. How much did each watch cost?

Answer: ....

PROBLEM 62

There were 24 potatoes in a box. Thomas bought 2 boxes and used 12 potatoes. He then repacked the remaining potatoes into packets of 6. How many packets of potatoes did Thomas pack?

Answer: ....

PROBLEM 63

There were 8 charity tickets in a booklet.
Linda sold 18 booklets and Jennifer sold
4 times as many booklets as her. Michael
sold half the number of booklets that Linda
sold. How many charity tickets did they sell
altogether?

Answer: ....

PROBLEM 64

At a sports shop, Larry bought 9 tennis rackets and 3 golf clubs for a total of $3300. Each golf club cost $350. How much did each tennis racket cost?

Answer: ....

PROBLEM 65

Grade 4 has 28 hours of study lessons each week. ¼ of the lessons are English.

¼ of the lessons are Math. How many lessons are NOT English or Math?

Answer: ....

PROBLEM 66

There are 64 children in Grade 4 . ½ of them have their lunch at school. ½ of the children having lunch at school choose pasta on Monday. How many children do not choose pasta on Monday?

Answer: ………………………..

PROBLEM 67

There are 81 children in Grade 4. A third of them take the school bus to come to school. How many children do NOT take the school bus?

Answer: ....

PROBLEM 68

There are 716 children in the whole school. ¼ of them are in kindergarten. ½ of them are in Primary and the rest are in Secondary. How many children are there in Secondary?

Answer: ....

74

There are 24 hours in a day and my parents
tell me that I should sleep for 2/3 of the day.
How much time should I spend on sleeping?

Answer: ....

Ben and George collected 225 stamps
together. Ben collected 3/5 of this amount.
How many stamps did George collect?

Answer: ....

PROBLEM 71

George is 184 centimeters tall. His sister Sarah is 5/8 as tall as him. How tall is Sarah? By how many centimeters is George taller than Sarah?

Answer: ....

PROBLEM 72

Sam baked some cupcakes. He gave 1/3 of them away and ate 1/12 of them. What fraction of the cupcakes were left?

Answer: ....

In a sports shop, a pair of roller skates cost $46 . The shopkeeper offered me a discount and said if I buy one pair, I could buy another pair for only 5/9 of the normal price. How much would the two pairs of roller skates cost me?

Answer: ....

Jimmy collected 256 marbles on Monday and 120 on Tuesday. He lost 3/4 of the total marbles on his way to school on Wednesday .How many marbles did Jimmy have left when he arrived at school on Wednesday?

Answer: …………………………

There are 40 students in the class. 3/5 of the students support Chelsea, 1/5 support Hull, the remainder support Arsenal. How many students support each team?

Answer: ....

PROBLEM 76

A candy shop normally sells Dairy Milk chocolate bars for 48 cents each. The shopkeeper says if I buy 3, I can have them for ¼ less than the normal price.

i) For how much money can I buy 3 Dairy Milk chocolate bars for?

ii) How much will each Dairy Milk chocolate bar cost if i buy 3?

Answer: ....

PROBLEM 77

Last year, Mrs. Rogers weighed 80 kg. This year she weighs 2/5 more. How much does Mrs. Rogers weigh this year?

Answer: ....

McDonalds sells milkshakes in two sizes. The small milkshake contains 300 ml and the large milkshake contains 2/5 more.

a) How much does a large milkshake contain?

b If Mr. Smith drank 2/3 of the small milkshake and Mr. Jacob drank 1/2 of the large Milkshake; who drank more?

Answer: ....

3/4 of the tickets for the Lady Gaga concert have been sold. If there are 4000 tickets on sale, how many tickets have been sold?

Answer: …………………………

PROBLEM 80

A newspaper vendor sold 2/3 of all his newspapers one morning. The rest were returned. If he had 390 newspapers;

a) How many newspapers did he sell?

b) How many newspapers did he return?

Answer: ....

PROBLEM 81

1/3 of the people in a room are under 30 years of age. If there are 25 people under 30, how many people are over 30?

Answer: ....

PROBLEM 82

Answer the following questions about fractions.

a) How many halves make up one whole?

b) How many quarters make up one half?

c) How many sixths make up one whole?

d) How many eights make up three quarters?

e) How many eights are same as three sixths?

Answer: ....

PROBLEM 83

There are 56 children in Grade 4. They are split in half into two classrooms.

a) How many children are there in each classroom?

b) If three quarters of all the children go on a school trip, what is the number of children who go?

c) What fraction of the children are left in the school?

Answer: ....

PROBLEM 84

There were three boxes of fruits: melons, mangoes and apples. The weight of mangoes was 1/2 kg less than the weight of melons. The weight of apples was 1/8 kg more than the weight of mangoes. If the weight of melons was 6 ¾ kg, what was the weight of the apples?

Answer: ....

At a football match, there were three times as many men as women and twice as many men as boys. 1/3 of the children were girls. If there were 120 girls;

a) How many boys were there?

b) How many adults were there?

Answer: ....

PROBLEM 86

Mia had 1/5 of 100 dollars. Cindy had 1/4 of 80 dollars. Who had more money?

Answer: ....

There are 75 children in Grade 4. A third (1/3) of them have brown hair. A third (1/3) of them have black hair. How many children do NOT have brown or black hair?

Answer: ....

PROBLEM 88

Joanna baked an apple pie. She ate 1/2 of the pie, Julie ate 4/8 of the pie, and Mark ate 25/100 of the pie. Is that possible? Why/why not?

Answer: ....

PROBLEM 89

Sam bought 1 foot long subway sandwich and cut it into 2 equal pieces.

He then ate one third of one part and one half of the other part.

How many inches of the sandwich was left?

Answer: ....

PROBLEM 90

There were 182 guests at Park Royal Hotel in Beijing. The guests wanted to take a taxi to the Great wall of China. Transport was provided at a fixed price of $27 for 4 guests . What was the lowest possible transport cost for these 182 guests?

Answer: ....

PROBLEM 91

To encourage David to work harder in
math, his mother said she would pay him
10 cents for each right answer and subtract
5 cents for each wrong answer.
If David earned 20 cents after doing 32
problems, how many problems did David
get right?
How many did he get wrong?
How many questions more would he need
to get right to earn one dollar?

Answer: ....

Tim received $14 to feed a neighbor's cat for 2 days. At this rate, how many days would he have to feed the cat to earn $40? The neighbor's family is going on vacation for 4 weeks next summer. Tim wants to earn enough money to buy a guitar that costs $90. Would he be able to earn enough money in 4 weeks?

Answer: ....

PROBLEM 93

Henry bought 4 books. He had 10 ten-dollar notes. If 2 books cost $7.90, how much money did Henry have left?

Answer: ....

PROBLEM 94

It costs $5.80 for adults to go to cinema and $3.40 for children. How many children did 4 adults take along with them if they spent $36.80?

Answer: ....

100

PROBLEM 95

Each month I visit a dentist to get my check-up done. I save $102 every year to cover the cost of my all my check-ups. If each visit to the dentist costs $10.75 will I be able to go for check up every month? If not, for how many months can I afford to go?

Answer: ....

PROBLEM 96

Katie buys a new book every week out of her $20 pocket money and it costs her $11.90 every time. She has not bought one for a while and has 5 weeks of pocket money saved up. However, she spent $10.90 to buy a birthday present for her sister. How many books can Katie buy when she goes shopping this time?

Answer: ....

PROBLEM 97

Following are the bus fares for adults and children:

Child single = 50 cents
Child return = 79 cents
Adult single = 80 cents
Adult return = $1.02

If I take my 7 children to town 7 times a year and buy all return tickets, how much will I spend on bus travel in an year?

Answer: ....

PROBLEM 98

Sandra bought 5 similar toy cars for $80. She also bought a doll which cost thrice as much as 1 toy car. How much did the doll cost?

Answer: ....

104

PROBLEM 99

If you bought 5 kilos of potatoes at 33 cents a kilo and a bag of onions for 29 cents a bag, how much change would you get back if you paid by a $5 note?

Answer: ....

105

PROBLEM 100

I went to the market and bought some groceries for my mother which cost $39.45. My mother had given me $60c to spend. How much change did I get back after paying for the groceries?

Answer: …………………………

PROBLEM **101**

Which is a better buy to save more money; to buy 5 packets of biscuits each with 10 biscuits inside and costing 50 cents per pack or buying 1 large packet with 110 biscuits costing $5.50?

Answer: ....

PROBLEM 102

Two mechanics worked on a car for 3 hours. First mechanic's labor cost is $6 per hour, the other mechanic's labor cost is $4.50 per hour. What is the total bill for labor?

Answer: ....

PROBLEM 103

Petrol costs $3 per liter. How much does it cost if I travel 30 km and my car uses 0.45 liters every 10 km?

Answer: ....

PROBLEM 104

Brian gave $3 to Alex. Alex then gave $5 to Brian. Brian now has $18 less than Alex. If Brian has $41 now, how much did Alex have at first?

Answer: …………………………

PROBLEM **105**

Jim and Jake are brothers and they each had the same amount of money which they put together to buy a toy. The cost of the toy was $22. If the cashier gave them a change of $6, how much money did each have?

Answer: ....

PROBLEM 106

Richard earns $15 an hour on weekdays
and two times more every two hours on
weekends. Richard works for 8 hours
each day on weekdays and 4 hours each
on Saturdays and Sundays. How much
will he earn in 1 week?

Answer: ………………………….

PROBLEM 107

There were 12 necklaces with 5 pearls each.
Jenny paid $20 for each pearl. How much did
she pay for all the necklaces?

Answer: ....

PROBLEM 108

Kevin bought four towels for $8 each and a blanket for $18. After he paid the shopkeeper handed him $50 back as change. What denomination was the note that Kevin used to pay?

Answer: ....

114

One mango drink costs $2.20, a strawberry drink costs $0.10 less than the mango drink and a peach drink costs $0.20 more than the mango drink. What is your total bill if you buy one of all three?

Answer: ....

PROBLEM 110

Jake's monthly grocery bill is $58. Jake said that with the money he earned during his summer job, he could buy groceries for two months, spend $130 to buy a guitar and still have half his money left. How much did Jake earn during his summer job?

Answer: ....

PROBLEM 111

Regina found a nice skirt for herself for $12.80, another one for $2.50 lesser, and yet another for $1.50 lesser.

If she buys all the three skirts, how much would be her total bill?

Answer: ....

PROBLEM 112

Liz, Beth, and Emma decided to equally share the cost of lunch they had together. Liz had spent $18, Beth had spent $15 and Emma had spent $6 while paying for the bill. How much did each person pay the others so that everyone spent the same amount?

Answer: ....

PROBLEM 113

Claire has $15.40 all in 20 cent coins. She puts them into stacks of three. How many stacks would she be able to make and how many cents would be left over?

Answer: ....

One ticket for the zoo costs 50 cents, 100 cents for the second ticket, 150 cents for the third ticket and so on. If you bought 10 tickets, how much would you pay altogether?

Answer: …………………………

PROBLEM 115

Tim, Jane, Laura, and Katie collect some money to buy a birthday present for their friend. Tim gave twice as much as Katie. Jane gave six dollars lesser than Katie. Laura gave 25 dollars more than Katie. Katie gave $75. How much did each child give?

Answer: ....

PROBLEM 116

Bernard earned $12000 each month. He spent $1500 on living expenses and saved 3 times that amount. Whatever money was left, he gave to his motherat the end of the year. How much money did he give to his mother in a year?

Answer: …………………………

Emma bought 6 pencil-boxes at $2.45 each.
How much change would she receive if she
paid $30 to the cashier ?

Answer: ....

PROBLEM **118**

Mr. Thomas had $35000. He donated
$7550 to a charity and divided the
remainder equally among his 6 sons. How
much did each son receive?

Answer: ....

PROBLEM 119

Cherie earns $42 an hour. She worked 1 hour and 45 minutes yesterday and 4 hours and 35 minutes today. How much did Cherie earn after working today and yesterday?

Answer: ....

125

PROBLEM 120

Rudy received $24.75 for his birthday. He wants to to save this money for three things: movie tickets, art supplies and books. If he wants to have the same amount of money for each thing, how much money will he put for each?

Answer: ....

126

PROBLEM 121

Jane wants to call Peter. Jane lives in Singapore and Peter lives in London. The clocks below show the time difference between Singapore and London. When it is 6.45 a.m. in Singapore, it is 11.45 p.m. in London. What time in Singapore should Jane call Peter if she wants to speak to him at 8 a.m. London time?

Singapore 1st Nov

London 31st Oct

Answer: ....

PROBLEM 122

Mike arrived at a party 29 minutes before 3:23 p.m. What time was it 15 minutes after he arrived?

Answer: ....

PROBLEM 123

It takes 55 minutes to get from Peter's house to the school. His school starts at 9.00 a.m.. It takes Peter 30 minutes to take his shower, 45 minutes to eat his breakfast. What time should Peter start getting ready to be on time for school?

Answer: ....

129

PROBLEM 124

Sandra woke up at 6:45 a.m. to get ready for school. She had slept for 8 and 1/4 hours. What time did she go to bed?

Answer: ....

PROBLEM 125

10 points are evenly marked on a race track. It took a runner 60 seconds to get from the first point to the third point. If the runner continues at the same speed, how long will it take him to run the complete track?

Answer: ....

PROBLEM 126

Claire is going to watch a movie. The movie starts at 6:30 p.m. Claire takes thirty-five minutes to take shower and another 45 minutes to get ready. It takes thirty five minutes to get from her house to the movie hall. At what time should she start getting ready?

Answer: ....

PROBLEM 127

Janet has to be at work by 9:00 a.m. and it takes her 45 minutes to get dressed, 20 minutes to eat her breakfast and 35 minutes to walk to work. What time should she get up?

Answer: ....

PROBLEM 128

Liz thinks it will take 2 hour and 15 minutes to do all her homework. She wants to watch a TV show at 8:00 p.m. What time should she start doing her homework to finish before the TV show?

Answer: ....

134

PROBLEM 129

Roger wants to attend a concert that starts at 5:00 p.m. He would need 25 minutes to get ready, 35 minutes to eat and 25 minutes to drive to the concert hall. He wants to arrive 10 minutes early. At what time should he start to get ready?

Answer: ....

PROBLEM 130

You should be in your classroom by 8:30 a.m. each school day. It takes you forty five minutes to take shower and get dressed, twenty minutes to eat breakfast and sixteen minutes to walk to school. What time should you get out of bed to get ready for school ?

Answer: …………………………

136

PROBLEM 131

Jason has to be at school by 8:30 a.m. It takes him 20 minutes to get dressed, 20 minutes to eat and 30 minutes to walk to school. What time should he get up?

Answer: ....

137

PROBLEM 132

Peter arrived at work at 15:45 today. It took him 90 minutes to get there from his home. At what time did he leave home?

Answer: ....

138

PROBLEM **133**

Gavin went for a walk at 10.25 a.m. He walked for 2 hours and 37 minutes. What time did he return home?

Answer: ....

PROBLEM 134

It takes a newspaper vendor 2 hour and 46 minutes to deliver his newspapers. He starts at 7.55 a.m. in the morning. What time does he get home?

Answer: …………………………

PROBLEM 135

In the last four days, Tim watched six TV programs each day. If each program was 35 minutes long, how much time in hours and minutes did Tim spend watching TV?

Answer: ....

PROBLEM 136

Brad spent 37 minutes on each worksheet he completed. He completed 17 worksheets. How much time in hours and minutes did Brad spend working on worksheets?

Answer: ....

PROBLEM 137

Kate took 24 minutes to finish the first 12 questions of an exam that has 50 questions in all. At this rate, how much more time will Kate need to finish the exam?

Answer: ....

143

PROBLEM 138

Lisa spent seventy-four minutes preparing for an exam. Jane spent 3 hours and 54 minutes preparing for the exam. How much longer did Jane spend preparing for the exam?

Answer: ……………………………

PROBLEM 139

Lily worked out on the treadmill for 55 minutes and then went cycling for 1 hour and 25 minutes. How long did Lily exercise?

Answer: ....

PROBLEM 140

Suzanne was baking a cake for her mother. It took thirty four minutes to prepare the batter. Then the cake baked in the oven for forty-five minutes. If Suzanne started baking at 12:43 p.m., by what time was the cake done?

Answer: ………………………

Arnold's flight was scheduled for departure at 9:44 a.m. The flight got delayed By ninety minutes. What is the new departure time?

Answer: ....

PROBLEM 142

Tina takes 18 minutes to finish each worksheet. Today she started working at 3:15 p.m. and completed three worksheets. What time did she finish?

Answer: ....

148

PROBLEM 143

Betty is at the airport. Her flight takes off at 11:17 p.m. The expected flight time is 4 hours and 45 minutes. What time would she land?

Answer: ....

Jack can get to school faster by cycling. He reaches school by 8:45 a.m. when he cycles and by 10:20 a.m. when he walks. How much quicker is it for Jack to cycle than to walk?

Answer: ....

PROBLEM 145

Jane is going to meet Paul at the movie theatre for the 10:00 p.m. show. If the current time is 5:15 p.m. how much time does Jane have until the show?

Answer: ....

151

PROBLEM 146

Siya is flying from New York to San Francisco with a stop over at Chicago. The plane will land in Chicago at 1:38 p.m. and will take off for San Francisco at 3:20 p.m. For how long would Siya be at Chicago?

Answer: ....

PROBLEM 147

Amie was invited to Mark's birthday party. The party started at 4:00 p.m. Amie reached sixty-five minutes after the party started. What time did Amie reach?

Answer: ....

PROBLEM 148

Jenny hasn't decided if she wants to see a movie at 5:00 p.m. or at 6:15 p.m. The time currently is 3:30 p.m. How much longer would Jenny need to wait for the second show compared to how long she would wait for the first show?

Answer: ....

154

PROBLEM 149

Paul decides to find out about his family and their history. He gets the following information from his parents. He decides to find out how old each of his ancestors were when they died.

	Born	Died
Mary	20/11/1942	11/01/2001
Benjamin	13/05/1954	04/11/1990
Frank	01/02/1851	13/12/1915
Martha	28/07/1934	11/03/2002
Sandra	30/06/1871	27/06/1954

a) List the family members in order of how long they lived.

b) List the family members in order of when they were born.

c) List the family members in order of when they died.

Answer: ....

155

PROBLEM 150

a) I am a unit of time on Earth. I am 1/7 of a week. What am I?

b) I am a unit of time. I am 1/60 of 1/60 of 1/24 of the answer to number one.

What am I?

Answer: …………………………

156

William is 7 times as old as his son this year.
His son was 4 years old last year. How old
will his son be when William is 60 years old?

Answer: ....

PROBLEM 152

Joshua is 12 years old now. His father is 32 years older than him. What will be the total of their ages in 5 years time?

Answer: …………………………

PROBLEM 153

Ben, Henry, John, and Peter had a running challenge among them. Henry ran twice as far as John. Ben ran 10 miles more than Henry. Peter ran eight miles less than Henry. John ran 25 miles. How many miles did each person run?

Answer: ....

PROBLEM 154

Laura weighs 14 kg. Her first sister is 4 times heavier than Laura. Her second sister is half the weight of her first sister. What is the weight of her 2 sisters put together?

Answer: …………………………

PROBLEM 155

Alvi has a bottle with a capacity of 726 ml. She poured 9 such bottles of water into an empty jug. However, 3005 ml of water overflowed from the jug. What is the capacity of the jug?

Answer: ....

PROBLEM 156

Tia made a sandwich with some cheese and 2 slices of bread that each weighed 2 grams. She weighed the sandwich and found it weighed 9 grams. How much cheese did she use?

Answer: ....

PROBLEM 157

A 2.5 litre bottle of cola is shared between 5 friends, how much does each person get?

Answer: ....

163

PROBLEM **158**

Michael drinks a 350 ml bottle of lemonade every day. How much lemonade will he drink in one week? How much is this in litres?

Answer: ....

164

PROBLEM 159

A car uses 3.5 liters of fuel every 2 kilometers it travels. How much fuel does it use if it travels 75 kilometers?

Answer: ....

165

PROBLEM 160

Amie has a jug of lemonade. She does not know how much lemonade she has, but she knows she can fill 12 glasses which have a capacity of 280 ml each. How much lemonade does Amie have?

Answer: …………………………

PROBLEM 161

Jug A holds 1800 ml water. Jug B holds 3/4 more. How much water does jug B hold? How much water do the two jugs hold?

Answer: …………………………

PROBLEM 162

Sarah creates a fruit punch. It contains 1/10 of a liter of apple juice, 2/5 of a liter of orange juice and 1/8 of a liter of grape juice. Which jug is the most suitable for Sarah to serve her fruit punch in?

| Jug 1 | Jug 2 | Jug 3 |
| 0.4 liters | 5 liters | 755 ml |

Answer: ....

168

Rick is 98 centimeters tall. Ted is 31 centimeters tall. How much taller in meters and centimeters is Rick than Ted ?

Answer: ………………………

PROBLEM 164

A piece of wood is 5 meters long.
A carpenter cuts off 45 centimeters. How long is the wood now in meters and in centimeters?

Answer: ....

Andy cuts a long pipe into 5 pieces. Each piece is 45 centimeters. What was the original length of the pipe in meters?

Answer: …………………………

Richard has 40 books. He needs 1.2 meters of shelving to store them. How wide is each book?

Answer: ....

PROBLEM 167

2.4 meters of ribbon is cut into 4 equal pieces. How long is each piece in meters and centimeters?

Answer: ....

PROBLEM 168

A carpenter has a plank of wood which is 3 meters long. He cuts it into 6 pieces. What is the length of each piece of wood in meters and centimeters?

Answer: ………………………….

PROBLEM 169

Mike participated in a 3000 meters race. He stopped after 1100 meters. How far is Mike still left to go to complete the race?

Answer: ....

175

PROBLEM 170

Paul dug 9 rows of earth to plant vegetables. Each row was 4 meters long. How much length of the ground did Paul dig altogether in meters and centimeters?

Answer: ....

PROBLEM 171

Julian and her friends wanted to bake a birthday cake for their friend. They needed 3.5 kg of sugar for the recipe. Sarah had 1.6 kg of sugar and Tim had 1300 g. How much more sugar did they need to buy to have enough for the recipe?

Answer: ....

PROBLEM 172

Peter needed 6 kg of sand for his cement mix. He only had 1.8 kg. While carrying the sack of sand he dropped and spilled another 0.9 kg. How much more sand does Peter need to have enough for making the cement mix?

Answer: ....

178

PROBLEM 173

Sandra wanted to post a parcel to her friend Kathy. Her parcel contained a 200 grams bar of chocolate, a 350 grams bag of sweets, 2 packets of biscuits that weighed 20 grams each and a cake that weighed 950 grams. Sandra forgot the weight of the parcel box. The entire parcel was weighed at the post office and it weighed 1560 grams in all. How much did the box weigh?

Answer: ....

PROBLEM 174

Bella bought a 3.2 kg fruit cake. Shane came in and cut the cake into half and took away one half to share with his friends. Then Paul came in and cut himself a slice. Bella then decided to weigh the cake to see how much was left and found that it weighed 900 grams now.

a) How much did Paul's piece weigh?

b) How much cake did Shane share with his friends?

Answer: ....

180

Laura bought a 2.8 kg jar of peanut butter for her Mum. She took out 700 grams of peanut butter separately from the jar for her friend. She bought 3 more small jars of peanut butter to refill the bigger jar for her mother. Each small jar weighed 225 g. What was the weight of the big jar after it was refilled?

Answer: ....

PROBLEM 176

Jenny drew a picture on a piece of paper and then put a border of 4 cm around the picture. The length of the paper was 24 cm. If the breadth of the paper was 8 cm, find the area of the border.

Answer: ………………………….

4 small squares are used to form a large square. If each side of a small square is 8 cm, what is the perimeter of the large square?

Answer: ....

PROBLEM 178

What will be the labor charge for tiling a hall 22 meters long and 17 meters wide at the rate of $8 per square meter?

Answer: ....

PROBLEM 179

Find the cost of painting a square board of sides 36 meters at the rate of $4 per square meter.

Answer: ....

PROBLEM 180

Ms. Jane wants to tile her room that is 12 feet wide and 25 feet long.

How many square tiles that are 1 foot on each side would be required to cover the floor?

How many tiles of dimensions 3' x 3' inches would be required to cover the same floor?

Answer: ....

The rectangle and the square shown below have the same perimeter. What is the length of one side of the square?

25cm

15cm

Answer: ....

187

Some 2 x 2 inches square cards are cut out from a square sheet that measures 2 feet by 2 feet. What is the greatest number of cards that can be cut from the sheet?

Answer: ……………………………

The figure below is made up of 6 squares. The side of each square is 4 cm. What is the perimeter of this figure?

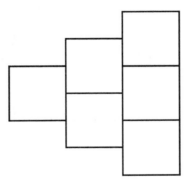

Answer: ....

PROBLEM 184

The length of the rectangle is thrice its breadth. What is the area of the rectangle?

Area = ?   9 cm

Answer: ………………………………

PROBLEM 185

If the length of the side of each little square is 1 cm, what is the area of letter N shown below?

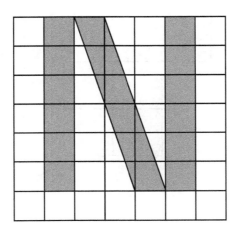

Answer: ....

The area of each of the squares below is 16 cm^2. Find the perimeter of the figure.

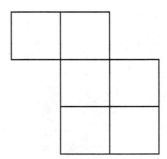

Answer: ....

The figure below is made up of 10 identical squares each having an area of 4 cm^2. Find the area of the whole figure.

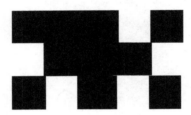

Answer: ....

Liz has a regular octagonal shaped box. She wants to decorate it with a ribbon all around the edges. Each side of the box measures 7.54 centimeters. How much ribbon does she need altogether?

Answer: ....

PROBLEM 189

Jimmy has been asked to put a fence all around a regular hexagonal garden. The perimeter of the garden is 46.8 meters. How much does each side measure?

Answer: ....

Cindy had a regular pentagon shaped box. She decorated it all around the edge with shiny circles. Each circle had a diameter of 2 centimeters. The perimeter of the box was 200 centimeters.

a) How many shiny circles did Cindy need altogether?

b) How many circles did she put on each side?

Answer: ....

PROBLEM 191

Farmer Tom has a square field. Its each side measures 120 meters. He wants to put a barbed wiring all around it.

a) How much wiring does he need?

b) Tom has another rectangular field measuring 15 meters by 21 meters. What is the perimeter of this field?

120 m

21 m

15 m

Answer: ....

197

PROBLEM 192

A rectangle has an area of 56 cm^2. The lengths of its sides are WHOLE numbers.

a) What is the smallest possible perimeter of this rectangle?

b) What is the largest possible perimeter of this rectangle?

Answer: ....

PROBLEM 193

Judie wants to have a new carpeting for her bedroom. Her bedroom is a 11 meters by 9 meters rectangle. If carpeting costs $6 per square meter, how much will it cost to carpet her entire bedroom?

Answer: ....

PROBLEM 194

Siya is making a display board for her school. The display board is a 15 feet by 9 feet rectangle. She needs to put a ribbon border around the entire display board. What is the length of the ribbon she needs? Express your answer in centimeters.

Answer: ....

PROBLEM **195**

Jasmine is making a poster for the school talent show. The poster is a 0.75 meters by 0.90 meters rectangle. If ribbon costs $2 per meter, how much will it cost to add a ribbon border around the entire poster?

Answer: ....

PROBLEM 196

Danny has a rectangular rose garden that measures 800 centimeters by 1200 centimeters. He wants to put fertilizer in his entire garden. One bag of fertilizer can cover 16 square meters. How many bags would he need to cover the entire garden?

Answer: ....

Ella wants to buy wood to make a frame for her picture. Her picture is a 12 feet by 10 feet rectangle. What is the total length of the wood strips in centimeters that she would need for her picture?

Answer: ....

PROBLEM 198

A side wall of my bedroom is 13 feet by 9 feet. A can of paint covers 50 square feet of the wall. Will it be enough to paint the entire wall? Explain.

Answer: ....

204

PROBLEM 199

Alex wants to put tiles on his bathroom floor. Each tile measures 1 square foot. His bathroom measures 6 feet by 8 feet. How many tiles would be needed to cover the entire floor?

Answer: ....

PROBLEM 200

There are 60 books in a pile. Each book is 3 millimeters thick. How high is the pile in centimeters?

Answer: …………………………

 Solution to Question **1**

The number is less than 100. That means it is a 2 digit number.

Units digit is a 4
Tens digit is a multiple of 3

Possible tens digit numbers are 3, 6 and 9
That means the possible numbers are 34, 64 and 94
The sum of my digits is an odd number. So it cannot be 64.

So there are two possible answers: 34 and 94

 Solution to Question **2**

I am a two digit number.
If you round me to the nearest hundred, my value will increase by ten.
Greatest 2 digit number rounded to nearest hundred will be 100.
Therefore the number has to be 10 less than 100.
100 – 10 = 90. I am 90.

 Solution to Question **3**

3 x 3 - 3 - 3 - 3 = 0   3 - 3 ÷ 3 - 3 ÷ 3 = 1   3 x 3 ÷ 3 - 3 ÷ 3 = 2
3 + 3 - 3 + 3 - 3 = 3   3 ÷ 3 + 3 ÷ 3 + 3 = 5   3 x 3 + 3 - 3 - 3 = 6
3 x 3 + 3 ÷ 3 - 3 = 7   3 x 3 ÷ 3 + 3 + 3 = 9

 Solution to Question **4**

Next reading when all digits are different will be 187590
No. of kilometers after which this will happen again = 187590 – 187569 =
21 km

 Solution to Question **5**

I am an even number between 5 and 20.
= 6, 8, 10, 12, 14, 16, 18
If you keep subtracting 5 from me, you will be left with 0.
i.e. it must be a multiple of 5
The number is 10

208

Solution to Question 6

The secret number has to be divisible by 3, 5 and 9. The LCM of 3, 5 and 9 is 45.
The multiple of 45 that lies between 275 and 325 is 315.
The secret number is 315.

Solution to Question 7

Smallest number = 2 x 3 + 6 = 6 + 6 = 12
Largest number = 6 x 3 + 2 = 18 + 2 = 20

Solution to Question 8

The number of coins has to be a multiple of 2 as well as 5. The LCM of 2 and 5 is 10.
The number of coins is between 30 and 50. As 40 is the only multiple of 10
between 30 and 40, the number of coins is 40.

Solution to Question 9

The sum of the digits is 10.
The number is an odd number and is less than 20.
The number is 19.

Solution to Question 10

Find a pair of numbers with:
a) a sum of 11 and a product of 24.
24 = 2 x 12
= 3 x 8
= 4 x 6
Out of these combinations only 3 + 8 = 11. Therefore the pair of numbers that
gives a sum of 11 and product of 24 is 8 and 3

b) sum of 40 and a product of 400.
400 = 20 x 20
= 10 x 40
= 8 x 50
= 5 x 80
= 4 x 100

209

Out of these combinations only 20 + 20 = 40. Therefore the pair of numbers that gives a sum of 40 and product of 400 is 20 and 20.

c) a sum of 15 and a product of 54
54 = 18 x 3
   = 2 x 27
= 6 x 9
Out of these combinations only 6 + 9 = 15. Therefore the pair of numbers that gives a sum of 15 and product of 54 is 6 and 9.

 ## Solution to Question 11

I am a number between 190 and 207.
Multiples of 4 between 190 & 207 are
192, 196, 200, 204
When divided by 4, the remainder is 2
Therefore the number can be above multiples + 2 =
194, 198, 202, 206

Multiples of 5 between 190 & 207
195, 200, 205
When divided by 5, the remainder is 1.
Therefore the number can be the above multiples + 1
196, 201, 206
The common number in both the lists is 206
The number is 206.

 ## Solution to Question 12

The number of pets is greater than 195 but less than 300
It has to be divisible by 3, 9 and 10.
The LCM of 3, 9 and 10 is 90.
The multiple of 90 that is between 195 and 300 is 270.
So the number of pets = 270

 ## Solution to Question 13

The country with the smallest area = UK
The country with the largest area = Russia

210

The difference between the areas of Russia and China
 = 17,097,242 – 9,599,094
=7,498,148 sq. km

The total area of all countries listed above
= 9,628,091 + 17,097,242 + 9,599,094 +242,300 +3,287,293
=39,854,020 sq. km
Order these countries from the largest to the smallest
areas Russia, USA, China, India, UK

Solution to Question 14

Pete starts at the bottom so he has to go up 24 steps.
He takes 4 steps per second so the time that Pete takes to reach the top =
24/ 4 = 6 seconds.
If they all have to finish in a tie, both Lisa and Amy should also take 6 seconds to reach the top.
Lisa takes 2 steps per second. So in 6 seconds she can cover 6 x 2 = 12 steps.
Amy takes 3 steps per second. So in 6 seconds she can cover 6 x 3 = 18 steps.

In order to end in a tie, Lisa should start at step number 12 (24 – 12 = 12) and Amy should start at step number 6 (24 – 18 = 6)

Solution to Question 15

Every 4th child is wearing spectacles.
Every 3rd child is girl.
Every 2nd child is wearing a white shirt.

12th child: 12 is a multiple of 2,3,4

So the child is a girl who wore a white shirt and spectacles

18th child: 18 is a multiple of 2 and 3
So the child is a girl who wore white shirt

21st child: 21 is a multiple of 3
So the child is a girl

211

Solution to Question 16

June 11th, 3 am.
30 hours earlier was 1 day and 6 hours earlier.
The clock will show June 11th, 3 a.m.

Solution to Question 17

As you observe the pattern, In pattern 1, we have 1 square
In pattern 2, we have 3 squares
In pattern 3, we have 5 squares..................so on

i.e. double the no. of squares – 1
So for 50th pattern the no. of squares = $2 \times 50 - 1 = 100 - 1 = 99$

Solution to Question 18

The sides that touch the table and the sides that touch the next cube don't get painted.
The top 8 faces will get painted.
Number of side faces that will get painted = 8 + 8 = 16 faces
The two faces at the front and end of the row of cubes will also get painted.
The number of faces painted = 8 + 16 + 2 = 26

Solution to Question 19

A street is 350 meters long.
Palm trees are planted on both sides of the street from beginning to the end of the street at 5 meters apart.
No. of palm trees on one side = 350/5 + 1 = 70 + 1 = 71
Total number of palm trees planted on both sides = 71 x 2 = 142

212

Solution to Question 20

If Aaron counts his stickers 2 at a time, he has 1 sticker left. That means he has odd number of stickers.

Possibilities are 101, 103, 105, 107 or 109

If he counts his stickers 5 at a time, he has 2 left. That means his number of stickers will be 2 more than a multiple of 5.

Possibilities are 102 and 107

The common number in both the groups is 107.

No. of stickers in Aaron's collection = 107

Solution to Question 21

To find the number let's solve the problem backwards and reverse all the mathematical operations that Claire did.
556 - 57 + 99 = 598
The number Claire thought of was 598.

Solution to Question 22

First week   2 chicken
Second week   4 chicken
Third week   8 chicken

The number of chicken doubles every week.

So at the end of the fourth week there will be 16 chicken and at the end of the fifth week there will be 32 chicken.

Solution to Question 23

Row 1   1 card
Row 2   4 cards
Row 3   7 cards

213

The number of cards increase by 3 in each row. Continuing this way,

Row 4	10 cards
Row 5	13 cards
Row 6	16 cards

 Solution to Question 24

No. of cuts	No. of pieces
1	2
2	4
3	8
4	16
5	32
6	64
7	128
8	256
9	512
10	1024

You will have 1024 (very tiny) pieces.

 Solution to Question 25

Russell has two dozen orange cookies = 2 x 12 = 24
He ate four cookies on Sunday
On every day that followed, he ate the number of cookies that was two greater than the day before.
On Tuesday he ate= 4 + 2 = 6
On Wednesday he ate = 6 + 2 = 8
By Wednesday total number of cookies he ate = 4 + 6 + 8 = 18
Remaining cookies = 24 − 18 = 6
Following the same pattern Russell should eat 8 + 2 = 10 cookies
On Thursday Russell will find that he doesn't have enough cookies to continue eating in the same pattern on Thursday.

 Solution to Question 26

There are 24 possible numbers that Jane can make:

214

3569	3596	3659	3695	3965	3956
5693	5639	5369	5396	5936	5963
6593	6539	6359	6395	6953	6935
9653	9635	9563	9536	9356	9365

The largest 4 digit number Jane can make = 9653

Solution to Question 27

Mike   Fred   Emma   Agnes   John

Solution to Question 28

LCM of 3,4 and 6 is 12
Therefore the day they all go to church on the same day again will be after 12 days

Solution to Question 29

A packet of plastic forks contains 10 forks.
A packet of plastic knives contains 12 knives.
LCM of 10, 12 is 60

The fewest number of packets you would have to buy to have exactly the same number of forks as knives are 6 packets of forks and 5 packets of knives.

Solution to Question 30

Nicole has 1 stamp left when she splits the stamps into groups of 2, 3 or 7.
This means the number of stamps will have to be one more than a multiple of 2, 3 and 7
The LCM of 2, 3 and 7 is 42. But the number of stamps is between 70 and 100. The next multiple of 42 is 84.
So the number of stamps = 84 + 1 = 85

215

Solution to Question 31

First person will shake 6 hands.
The second person will shake 5 hands (he has already shaken hands with the first person).
The third person will shake 4 hands (He has already shaken hands with the first and the second person so those wont be counted).
And so on.
So the number of hand shakes = 6 + 5 + 4 + 3 + 2 + 1 = 21

Solution to Question 32

Mr. David placed 17 fence posts 6 feet apart.

The distance from the first fence post to the last = (17 − 1) x 6 feet
(Note: There are 17 posts, but there are only 16 gaps between these posts)

= 16 x 6 feet
= 96 feet

Solut on to Quest on 33

Order of the players to finish the game is
Mary   Sandra   Jenny   Ben

Solution to Question 34

Let us write down the multiples of 7 and 8 in between 40 and 80
Multiples of 7: 42 49 56 63 70 77
Multiples of 8: 48 56 64 72

We have to now find a common number that is 1 more than a multiple of 8 and 2 more than a multiple of 7.

Such a number is 65
(65 − 2 = 63 is a multiple of 7 and 65 − 1 = 64 is a multiple of 8)

The number of chairs = 65

216

Solution to Question 35

a) 1 + 1 - 1 x 1 = 1
b) 5 x 5 x 5 ÷ 5 = 25

Solution to Question 36

Sam, Carla and Sarah spent one afternoon collecting sea shells. Sam collected
11. If we add the number of sea shells collected by Sam and Carla, the total
would be 24
11 + Carla's sea shells = 24
Carla's sea shells = 24 − 11 = 13
If we add the number of sea shells collected by Carla and Sarah, the total would
be 25 shells.
13 + Sarah's sea shells = 25
Sarah's sea shells = 25 − 13 = 12
Sea shells collected by Sam = 11
Sea shells collected by Carla = 13
Sea shells collected by Sarah = 12

Solution to Question 37

Lucy was 130 cm tall when she was 8 years old.
In the next year she grew 5 cm = 130 + 5 = 135 cm The
next year she grew 3 cm less than the previous year. So
she grew 5 − 3 = 2 cm
Her height at the age of 10 years = 135 +2 = 137 cm

Solution to Question 38

In each group, there were 18 boys and 8 fewer girls than boys.
Number of boys in each group = 18

217

Number of girls in each group = 18 – 8 = 10
Total number of students in each group = 18 + 10 = 28
Number of students who went for the school picnic = 28 x 5 = 140

Solution to Question 39

Tim caught 66 fish.
Bill caught 28 fish more than Alice and Alice caught 15 fish less than
Tim. Number of fish Alice caught = 66 – 15 = 51
Number of fish Bill caught = 51 + 28 = 79

Solution to Question 40

In a car park, there are 198 buses.
There were 54 more buses than vans.
Number of vans = 198 – 54 = 144
There are 36 more motorcycles than vans.
Number of motor cycles = 144 + 36 = 180
Total number of vehicles altogether = 198 + 144 + 180 = 522

Solution to Question 41

Mr. James bought 19 crates of fruits.
There were 64 fruits in each crate.
Total number of fruits = 19 x 64 = 1216
320 were apples, 414 were pears and the rest were mangoes
Total number of apples and pears = 320 + 414 = 734
No. of mangoes = 1216 – 734 = 482

Solution to Question 42

Sam, Hansel, and Amie had a collection of cards.
Sam had 260 cards.
Amie had twice as many cards as Sam = 260 x 2 = 520
Hansel had as many cards as Sam and Amie together = 520 + 260 =
780 Number of cards Hansel had = 780

218

Solution to Question 43

Row # 1 had 10 plants.
Row # 2 had twice as many plants as the first = 10 x 2 = 20
Row # 3 and Row # 5 had the sum of rows 1 and 2 = 10 + 20 = 30
Row # 4 and Row # 6 had the sum of rows 1,2, and 3 = 30 + 30 = 60 Number
of plants that Mark planted in his garden = 10 + 20 + 30 + 60 + 30 + 60 = 210

Solution to Question 44

Total number of pencils = 20 x 24 = 480
Pencils given away = 30 x 6 = 180
Number of pencils left with Ryan = 480 − 180 = 300

Solution to Question 45

Sarah went for the roller coaster ride 5 times altogether.
Sam went four more times than Sarah but three less times than Emily
= 5 + 4 = 9 times
No. of times Emily went = 9 + 3 = 12
Emily went twice as many times as Lucy.
No. of times Lucy went = 12/2 = 6
Lucy went 6 times for the roller coaster ride.

Solution to Question 46

18 people share a barrel of apples equally. Each person gets 12
apples. Number of apples = 18 x 12 = 216
If there were 6 fewer people, the number of people = 18 - 6 = 12
Number of apples each person would get = 216/12
= 18 apples

Solution to Question 47

A machine can print 70 cards in 6 minutes.
2 hours has 120 minutes = 120/6 = 20 groups of 6 minutes
Number of cards the machine can print in 2 hours = 70 x 20 = 1400

219

Solution to Question **48**

On Tuesday, Sam ordered 6 times as many cans of coke as he did on Monday
= 145 x 6 = 870
Number of cans of coke Sam ordered on Tuesday than on Monday
= 870 − 145
= 725

 Solution to Question **49**

Amy bought 95 boxes of mangoes. There were 50 mangoes in each box.
Total number of mangoes = 95 x 50 = 4750
Mangoes left after giving away to her friends = 4750 − 25 = 4725
Each new box contained 5 mangoes.
a) Number of small boxes Amy packed = 4725/5 = 945

Amount of money Amy earned if she sold each box for $15 = 945 x 15 = $14,175

 Solution to Question **50**

Bernard has sixteen pets.
All of his pets are cats and dogs.
Bernard has four more cats than dogs.
We'll take away the extra four cats and then divide the pets in equal number so that
number of cats remains four more than the number of dogs = 16 − 4 = 12 pets
Dividing 12 pets equally = 12/2 = 6
Number of cats Bernard has = 10
Number of dogs Bernard has = 6

 Solution to Question **51**

There are 6 large boxes.
Number of small boxes = 6 x 4 = 24
Number of smaller boxes = 24 x 3 = 72
Number of boxes altogether after counting all sizes
= 72 + 24 + 6 = 102

220

Solution to Question 52

Suzy has twenty eight dolls = 28 dolls
Sarah has thrice as many dolls as Suzy, so Sarah has = 28 x 3 = 84 dolls
Suzy has eight more dolls than Nicole, so Nicole has = 28 – 8 = 20 dolls
Amanda has four dolls less than Nicole, so Amie has = 20 – 4 = 16 dolls

Solution to Question 53

When wheel B rotates two times, wheel A rotates 6 times. 6 rotations of
wheel A = 2 rotations of wheel B
1 rotation of wheel A = 2/6 rotations of wheel B
40 rotations of wheel A = 2 ÷ 6 x 40 = 13 1/3 rotations of wheel B Wheel
B would rotate 13 full rotations.

Solution to Question 54

A box can hold 450 marbles.
A container can hold 5 times as many marbles as a box = 450 x 5 = 2250 marbles
No. of marbles 5 boxes and 2 containers can hold is
= 5 x 450 + 2 x 2250
= 2250 + 4500
= 6750 marbles

Solution to Question 55

A buffet lunch costs $23 for an adult and $7 less for a child
Cost of lunch for a child = 23 – 7 = $16
Total amount of money that a family of 2 adults and 3 children had to pay for
the buffet lunch
= 2 x 23 + 3 x 16
= 46 + 48
= $94

221

Solution to Question 56

Doris bought 8 packets of tomatoes and gave them to 25 students.
When she gave each of them 4 tomatoes, she had 4 tomatoes left. Total
tomatoes given to students = 25 x 4 = 100
Tomatoes left with her = 4
Total tomatoes = 100 + 4 = 104
Number of tomatoes in each packet = 104/8 = 13 tomatoes

Solution to Question 57

A shopkeeper sold 230 potatoes on Monday and twice as many potatoes
on Tuesday.
Number of potatoes sold on Tuesday = 2 x 230 = 460
He had 150 potatoes left.
Total number of potatoes = 460 + 230 + 150 = 840
He wanted to sell equal number of potatoes within 2 days.
Number of potatoes he should sell on each day = 840/2 = 420

Solution to Question 58

9 packets of sweets cost $72 and 3 packets of biscuits cost $12.
1 packet of sweets = 72/9 = $8
1 packet of biscuits = 12/3 = $4
Alice wanted to buy 4 packets of sweets and 2 packets of biscuits.
= 4 x 8 + 2 x 4
= 32 + 8
= $40
The total amount of money that Alice had to pay for the items = $40

Solution to Question 59

Florence bought 18 packets of apples weighing 5 kg each.
Total weight of apples = 18 x 5 = 90 kg
She then packed them into plastic bags of 2 kg each.
Number. of plastic bags she used = 90/2 = 45

222

Solution to Question 60

8 people paid $90 each for a dinner.
Amount of money paid by 8 people = 90 x 8 = $720
The organizers were still short of $24.
Actual amount that 8 people should have paid = 720 + 24 = $744
Actual amount of money that each of them should pay = 744/8 = $93

Solution to Question 61

Tracy bought 2 similar watches and had $150 left.
She spent 3 times the amount of money left with her.
Money spent = 150 x 3 = $450
Cost of each watch = 450/2 = $225

Solution to Question 62

There were 24 potatoes in a box.
Thomas bought 2 boxes
Number of potatoes in 2 boxes = 24 x 2 = 48
He used 12 potatoes.
Remaining potatoes = 48 − 12 = 36
He then repacked the remaining potatoes into packets of 6.
Number of packets of potatoes Thomas repacked = 36/6 = 6

Solution to Question 63

There were 8 charity tickets in a booklet.
Linda sold 18 booklets and Jennifer sold 4 times as many booklets as her.
Number of tickets Linda sold = 18 x 8 = 144
Number of tickets Jennifer sold = 4 x 18 x 8 = 576
Michael sold half the number of booklets that Linda sold.
= 18/2 x 8 = 72
Number of charity tickets they sold altogether = 144 + 576 + 72 = 792

223

Solution to Question **64**

Larry bought 9 Tennis rackets and 3 golf clubs for a total of $3300.
Each golf club cost $350.
Cost of 3 golf clubs = 3 x 350 = $1050
Cost of 9 Tennis rackets = 3300 − 1050 = $2250
Cost of 1 Tennis racket = 2250/9 = $250
Cost of each Tennis racket = $250

Solution to Question **65**

Grade 4 has 28 hours lessons each week.
1/4 of the lessons are English = 28/4 = 7
1/4 of the lessons are Math = 28/4 = 7
Total = 7 + 7 = 14
Number of lessons which are NOT English or Math = 28 - 14 = 14 lessons

Solution to Question **66**

In Grade 4 there are 64 children.
1/2 of them have their lunch at school = 64/2 = 32
1/2 of the children having lunch at school choose pasta on Monday = 32/2 = 16
Number of children who do not choose pasta on Monday = 32 − 16 =
16 students

Solution to Question **67**

There are 81 children in Grade 4
A third of them take the school bus to come to school = 81/3 = 27
Number of children who do NOT take the school bus = 81 − 27 = 54

Solution to Question **68**

There are 716 children in the whole school.
1/4 of them are in kindergarten = 716/4 = 179

224

1/2 of them are in Primary = 716/2 = 358
Total = 179 + 358 = 537
The rest are in Secondary = 716 − 537 = 179
Number of children in Secondary = 179

 Solution to Question **69**

There are 24 hours in a day and my parents tell me that I should sleep for 2/3 of the day.
Amount of time I should spend on sleeping = 2/3 x 24 = 16 hours

 Solution to Question **70**

Ben and George together collected 225 stamps together.
Ben collected 3/5 of this amount = 3/5 x 225 = 135
Number of stamps George collected = 225 − 135 = 90

 Solution to Question **71**

George is 184 cm tall.
His sister Sarah is 5/8 as tall as him = 5/8 x 184 = 115 cm
Sarah is 115 cm tall.
Difference in their heights = 184 − 115 = 69 cm
George is 69 cm taller than Sarah.

 Solution to Question **72**

1/3 + 1/12 = 4/12 + 1/12 = 5/12
Sam gave away and ate 5/12 of the cupcakes
1 - 5/12 = 7/12
7/12 of the cupcakes were left.

225

Solution to Question 73

A pair of roller skates costs $46 in a sports shop.
The shopkeeper offered me a discount and said if I buy one I can buy another pair for only 5/9 of the normal price.
Cost of the second pair = 5/9 x 46 = $25.56
Cost of the two pairs of roller skates = 46 + 25.56
= $71.56

Solution to Question 74

Jimmy collected 256 marbles on Monday and 120 on Tuesday
Total marbles = 256 + 120 = 376
He lost 3/4 of the total marbles on his way to school on Wednesday
= 3/4 x 376 = 282
When he arrived at school number of marbles Jimmy had left = 376 − 282 = 94

Solution to Question 75

There are 40 students in the class.
3/5 of the students support Chelsea = 3/5 x 40 = 24 students
1/5 support Hull = 1/5 x 40 = 8 students
Remaining support Arsenal = 40 − 24 − 8 = 40 − 32 = 8 students

Solution to Question 76

A candy shop normally sells Dairy Milk chocolate bars for 48 cents each.
Cost of 3 dairy milk chocolates = 48 x 3 = 144 cents
The shopkeeper says if I buy 3, I can have them for 1/4 less than the normal Price = 144 − 1/4 x 144
= 144 − 36
= 108 cents i.e. $ 1.08

a) 3 Dairy Milk chocolate bars will cost 108 cents.
b) Each Dairy Milk chocolate bar will cost = 108/3 = 36 cents

226

 **Solution to Question 77**

Last year, Mrs. Rogers weighed 80 kg.
This year she weighs 2/5 more
= 80 + 2/5 x 80
= 80 + 32
= 112 kg
Mrs. Rogers weight this year = 112 kg

 **Solution to Question 78**

McDonalds sells milkshakes in two sizes.
The small milkshake contains 300 ml and the large milkshake contains 2/5
more. Large milkshake = 300 + 2/5 x 300
= 300 + 120
= 420 ml

a) The large milkshake contains 420 ml.

b) Mr. Smith drank 2/3 of the small milkshake = 2/3 x 300 = 200 ml
Mr. Jacob drank 1/2 of the large milkshake= 1/2 x 420 = 210 ml
Mr. Jacob drank more.

 **Solution to Question 79**

3/4 of the tickets for the Lady Gaga concert have been sold.
There are 4000 tickets on sale.
Tickets that have been sold = 3/4 x 4000 = 3000

 **Solution to Question 80**

A newspaper vendor sold 2/3 of all his newspapers one
morning. The rest were returned. If he had 390 newspapers;

a) Number of newspapers he sold = 2/3 x 390 = 260
b) Number of newspapers he returned = 390 − 260 = 130

227

Solution to Question 81

1/3 of the people in a room are under 30 years of age and there are 25 people under 30.
1/3 of number of people in the room = 25
Number of people in the room = 25 x 3 = 75
Therefore number of people who are over 30 = 75 – 25 = 50

 Solution to Question 82

a) Number of halves that make up one whole = 1/2 + 1/2 = 1 2 halves make one whole
b) 1/4 + 1/4 = 1/2
Number of quarters (4ths) that make up one halve = 2
c) 1/6 + 1/6 + 1/6 + 1/6 + 1/6 + 1/6 = 6/6 = 1
Number of sixths that make up one whole = 6
d) 1/8 + 1/8 + 1/8 + 1/8 + 1/8 + 1/8 + = 6/8 = 3/4
Number of eights that make up three quarters = 6
e) 3/6 is same as 1/2
1/8 + 1/8 + 1/8 + 1/8 = 4/8 = 1/2
Number of eights that are same as three sixths = 4

 Solution to Question 83

There are 56 children in Grade 4. They are split in half into two classrooms.

a) Number of children in each classroom = 56/2 = 28
b) Three quarters of all the children go on a school trip = 3/4 x 56 = 42 children
c) Fraction of children left in school = 1 – 3/4 = 1/4

 Solution to Question 84

Weight of melons = 6 ¾ = 27/4 kg
Weight of mangoes was 1/2 kg less than the weight of melons = 27/4 – 1/2
= 25/4 kg
Weight of apples is 1/8 kg more than the weight of mangoes
= 25/4 + 1/8

228

= (50 + 1)/8
= 51/8 kg or 6.37 kg

 ## Solution to Question **85**

Number of girls = 120
1/3 of children were girls, so number of children = 360
Number of boys = 360 − 120 = 240.
Number of men is twice the number of boys = 240 x 2 = 480
There were three times as many men as women, so number of women = 480/3
= 160
Number of boys = 240
Number of adults = number of men + number of women = 480 + 160 = 640

 ## Solution to Question **86**

Amount of money Mia had = 1/5 x 100 = 20 dollars
Amount of money Claire had = 1/4 x 80 = 20 dollars
They both had equal amount of money.

 ## Solution to Question **87**

In Grade 4 there are 75 children.
A third (1/3) of them have brown hair = 75/3 = 25 students
A third (1/3) of them have black hair = 75/3 = 25 students
Total = 25 + 25 = 50
Number of children who do NOT have brown or black hair = 75 − 50 = 25
students

229

Solution to Question **88**

Joanna baked an apple pie.
She ate 1/2 of the pie.
Remaining pie = 1 − 1/2 = 1/2
Julie ate 4/8 of the pie = 1/2
Mark ate 25/100 of the pie = 1/4
It is not possible because there is no more apple pie left for Mark. Joanna and Julie ate one half each.

 ## Solution to Question **89**

1 foot = 12 inches
The length of each piece = 12/2 = 6 inches
Sam ate 1/3 of one = 1/3 x 6 = 2 inches. Remaining = 4 inches
He ate 1/2 of the other = 1/2 x 6 = 3 inches. Remaining = 3
inches Total sandwich remaining = 3 + 4 = 7 inches

 ## Solution to Question **90**

There were 182 guests at Park Royal Hotel in Beijing. The guests wanted to take a taxi to the Great wall of China.
Transport was provided at a fixed price of $27 for 4 guests.
Number of taxis required = 182/4 = 46
The lowest possible transport cost for these 182 guests = 46 x 27 = $1242

 ## Solution to Question **91**

To encourage David to work harder in math his mother said she would pay him 10 cents for each right answer and subtract 5 cents for each wrong answer.
He earned 20 cents after doing 32 problems.
Let the right answers be 'X'.
Wrong answers = 32 − X
X x 10 - (32 − X) x 5 = 20
10X - 160 + 5X = 20
15X = 20 + 160
X = 180/15 = 12

230

Number of problems David got right = 12
Number of problems he got wrong = 32 − 12 = 20
David has already earned 20 cents. To earn $1, amount of money David needs
to earn = 100 − 20 = 80 cents
Number of questions David needs to get right to earn one dollar = 80/10 = 8

Solution to Question 92

Tim received $14 to feed a neighbor's cat for 2 days.
Therefore in 1 day he earned = 14/2 = $7
At this pay rate, number of days he has to feed the cat to earn $40
= 40/7 = 5.7
= 6 days
The neighbor's family is going on vacation for 4 weeks next summer.

1 week = 7 days
Amount of money Tim can earn in 4 weeks = 4 x 7 x 7 = $196
Tim wants to earn enough money to buy a guitar that costs $90.
Yes, he would have enough money to buy the guitar.

Solution to Question 93

10 ten-dollars = $100
$7.90 + $7.90 = $15.80
$100 - $15.80 = $84.20
Henry had $84.20 left

Solution to Question 94

It costs $5.80 for adults to go to the cinema and $3.40 for children.
Let the number of children be 'X'
They spent $36.80
3.40X + 4 x 5.80 = 36.80
3.40X = 36.80 − 23.20
X = 13.60/3.40
X = 4
The 4 adults took 4 children along with them.

231

Solution to Question 95

Each month I visit a dentist to get my check-up done.
I save $102 every year to cover the cost of my check-ups.
Each visit to the dentist costs $10.75
For 1 year, the total cost to visit dentist = 10.75 x 12 = $129
No, my saving will not be sufficient to go every month.
The number of months I can afford to go = 102/10.75 = 9.48 =
9 months

Solution to Question 96

Katie buys a new book every week out of her $20 pocket money and it costs
$11.90 every time.
Katie's saving = 20 − 11.90 = $8.10
She has not bought one for a while and has 5 weeks of pocket money saved
up. Total money saved with Katie = 5 x 8.10
= $40.50
However, she spent $10.90 to buy a birthday present for her sister. Money
left with her = 40.50 - 10.90 = 29.60
Number of books Katie can buy when she goes shopping this time
= 29.60/11.90 = 2.48
= 2 books

Solution to Question 97

Child single = 50 cents. Child return = 79 cents. Adult single = 80 cents. Adult
return = $1.02.
 Cost of travelling once for all 7 children = 7 x 79 = 553 cents i.e. $ 5.53
Cost of travelling for children 7 times = $5.53 x 7 = $38.71
Amount I spend on myself to travel 7 times = 7 x $1.02 = $7.14
Total money spent = $38.71 + $7.14 = $45.85

Solution to Question 98

Cost of 1 toy car = $80/5 = $16
Cost of 1 doll = $16 x 3 = $48
The doll cost $48.

232

Solution to Question **99**

If you bought 5 kilos of potatoes at 33 cents a kilo and a bag of onions costing 29 cents a bag.
Total cost = 5 x 33 + 29 = 165 + 29 = 194 cents i.e. $1.94
Amount of change you would get from $5 = 5 − 1.94 = $3.06

 ## Solution to Question **100**

I went to the shop and bought some groceries for my Mum.
They cost $39.45.
Mum gave me $60.
Amount of change I got = 60 − 39.45 = $20.55

 ## Solution to Question **101**

Cost per biscuit while buying 5 packets at a price of 50 cents per packet =
50/10 = 5 cents
Cost per biscuit while buying one large packet with 110 biscuits costing $5.50
(550 cents)=
 550/110 = 5 cents
As the cost per biscuit is the same in both cases, both are equivalent buys.

 ## Solution to Question **102**

Two mechanics worked on a car for 3 hours.
First mechanics labor cost is $6 per hour, the other mechanics labor cost is $4.50 per hour.
The total bill for labor = 6 x 3 + 4.5 x 3
= 18 + 13.5 = $31.5

 ## Solution to Question **103**

Petrol costs $3 cents per liter.
I travel 30 km and my car uses 0.45 liters every 10 km
Total petrol used for 30 km = 0.45 x 30/10 = 1.35 liters
Cost of travel = 1.35 x 3 = $4.05

233

Solution to Question 104

Before Alex gave $5 to Brian, Alex had = 59 + 5 = $64
Before Brian gave $3 to Alex, Alex had = 64 - 3 = 61
Alex had $61 at first.

Solution to Question 105

Jim and Jake are brothers and they each had the same amount of money which they put together to buy a toy.
The cost of the toy was $22.
The cashier gave them a change of $6
Total amount = 22 + 6 = 28 $
Amount of money each had = 28/2 = $14

Solution to Question 106

Richard earns $15 an hour on weekdays. There are 5 weekdays in 1 week.

Amount of money earned on weekdays = 5 x 15 x 8 = 600

On weekends, he earns two times more every two hours and he works for 8 hours on Saturday and Sunday combined (4 hours on each weekend)

two times of $15 is $30.

Amount of money earned on weekends = 4 x 30 =120
Total amount of money earned by Richard = 600 + 120 = 720

Solution to Question 107

There were 12 necklaces with 5 pearls each.
Total number of pearls = 12 x 5 = 60

234

Jenny paid $20 for each pearl.

The amount she paid for all the necklaces = 60 x 20 = $ 1200

 **Solution to Question 108**

Kevin bought four towels for $8 each and a blanket for $18. Total cost = 4 x 8 +
18
= 32 + 18
= $50
After he paid the shopkeeper handed him $50 back as change.
Denomination of the note that Kevin used to pay was $100.

 **Solution to Question 109**

One mango drink costs $2.20
A strawberry drink costs $0.10 less than the mango drink.
Cost of a strawberry drink = 2.20 − 0.10 = $2.10
A peach drink costs $0.20 more than mango drink .
Cost of a peach drink = 2.20 + 0.20 = $2.40
Total bill if we buy one of all three = 2.20 + 2.10 + 2.40 = $6.70

 **Solution to Question 110**

Let the total earnings be X.
Jake's monthly grocery bill is $58.
Cost of buying groceries for two months = 2 x 58 = $116
Cost of guitar $130
After all his spending Jake would still have half his money left = X/2
So,
116 + 130 + X/2 = X
X - X/2 = 246
X/2 = 246
X = 246 x 2 = $492

Jake earned $492 during his summer job.

235

Solution to Question **111**

First skirt costs $12.80
Another for $2.50 less = 12.8 0 − 2.50 = $10.30
Yet another for $1.50 less = 12.80 − 1.50 = $11.30
 Cost to buy all three = 12.80 + 10.30 + 11.30 = $34.40

 ### Solution to Question **112**

Total money spent on lunch was 18 +15 + 6 = $39

If the cost was shared equally then each person should pay 39/3 = $13

 Emma needs to give Liz 18 -13 = $5

Emma needs to give Beth 15 − 13 = $2

 ### Solution to Question **113**

Number of 20 cents coins Claire has = $15.4 / 0.2 = 15.4 x 5 = 77 coins

If she puts them into stacks of 3, she would have 25 stacks and 2 coins would be left over.

 ### Solution to Question **114**

One ticket costs   50 cents
Two tickets cost 100 cents
Three tickets cost 150 cents

For every increase in ticket, the cost increases by 50 cents

If you bought 10 tickets, total cost
= 50 + 100 + 150 + 200 + 250 + 300 + 350 + 400 + 450 + 500
= 2750 cents

Amount to be paid altogether = $27.50

236

Solution to Question 115

Katie gave $75.
Tim gave twice as much as Katie = 75 x 2 = $150
Jane gave six dollars lesser than Katie = 75 – 6 = $69
Laura gave 25 dollars more than Katie = 75 + 25 = $100

 ## Solution to Question 116

Bernard earned $12000 each month.
He spent $1500 on living expenses and saved 3 times that amount.

Savings = 1500 x 3 = $4500

Total money used = $1500 + $4500 = $6000

Money left after expenses and savings given to mother =
$12000 - $6000 = $6000

Money given to mother in a year = $6000 x 12 = $72000

 ## Solution to Question 117

Emma bought 6 pencil-boxes at $2.45 each.
Total cost = 6 x 2.45 = $14.70

Change she would received if she paid $30 to the cashier $30 = 30 – 14.70 = $15.30

 ## Solution to Question 118

Mr. Thomas had $35000.
He donated $7550 to a charity.
Money remaining = 35000 – 7550 = $27450
He divided the remainder equally among his 6 sons.
Each son received = 27450/6 = $4575

 ## Solution to Question 119

Cherie earns $42 an hour.
She worked 1 hour and 45 minutes yesterday and 4 hours and 35 minutes today.

237

Total number of hours Cherie worked = 1:45 + 4:35 = 6 hours 20 minutes i.e. 6 and 1/3 hours
Cherie earns $42 an hour.
In 1 hour she earns $ 42
In 6 hours she will earn 6 x $ 42 = $252
In 1/3 hours, she will earn 42/3 = $14
Amount of money that Cherie earned for working today and yesterday = $252 + $14 = $266

 Solution to Question **120**

Rudy received $24.75 for his birthday.
He wants to to save this money for movie tickets, art supplies and books equally = 24.75/3 = 8.25
Amount of money Rudy will put for each = $8.25

 Solution to Question **121**

Difference in time between Singapore and London = 11: 45 – 6:45 = 7 hours

The time in Singapore when Jane should call Peter if she wants to speak to him at 8 a.m. London time = 8 + 7 = 15 i.e. 3 pm

 Solution to Question **122**

Time at which Mike arrived at the party = 3:23 – 0 : 29 = 2 : 54 p.m.

The time 15 minutes after he arrived = 2: 54 + 0:15 = 3: 09 p.m.

 Solution to Question **123**

It takes 55 minutes to get from Peter's house to the school.

238

It takes Peter 30 minutes to take his shower, 45 minutes to eat his breakfast.
Total no. of minutes Peter needs to get ready and reach school = 55 + 30 + 45 = 130 minutes = 2 hours 10 minutes
Counting 2 hours and 10 minutes backwards from 9.00 a.m. you get 6:50 a.m.

Therefore Peter should start getting ready at 6:50 a.m.

Solution to Question 124

Sandra woke up at 6:45 A.M.
She had slept for 8 and 1/4 hours = 8 hr 15 min

Time she went to bed = 6:45 − 8:15 = 10:30 P.M.

Solution to Question 125

Time taken to run each stretch between two points = 60/2 = 30 seconds
Remaining stretches to be completed = 7
The runner continues at the same speed, time taken by him to run the complete the remaining track = 7 x 30 = 210 seconds
Total time = 210 + 60 = 270 seconds

Solution to Question 126

The movie starts at 6:30 p.m.
Time Claire needs before the start of the movie =
35 minutes (shower + 45 minutes (getting ready + 35 minutes (to get to the hall
Total = 35 + 45 + 35 = 115 min i.e. 1 hour 55 min.

Time she should start getting ready = 6:30 − 1 hour 55 minutes = 4:35p.m.

Solution to Question 127

It takes Janet 45 minutes to get dressed, 20 minutes to eat her breakfast and 35 minutes to walk to work.
Total time needed = 45 + 20 + 35 = 100 min i.e. 1 hour 40 min

Time she should get up = 9 :00 − 1 hour 40 minutes = 7:20 a.m.

239

Solution to Question 128

Liz wants to watch a TV show at 8:00 p.m.
Time at which she should start doing her homework to finish before the
TV show = 8:00 – 2 hours 15 minutes
= 5:45 p.m.

Solution to Question 129

Roger would need 25 minutes to get ready, 35 minutes to eat and 25
minutes to drive to the concert hall.

He wants to arrive 10 minutes early.
Total time = 25 + 35 + 25 + 10 = 95 min i.e. 1 hour 35 min

Time at which he should start getting ready = 5:00 – 1 hour 35 minutes =
3:25 p.m.

Solution to Question 130

It takes you forty five minutes to take shower and get dressed, twenty minutes
to eat breakfast and sixteen minutes to walk to school.
Total = 45 + 20 + 16 = 81 = 1 hour 21 min

Time should you get out of bed to get ready for school
= 8:30 – 1 hour 21 minutes
= 7:09 a.m.

Solution to Question 131

It takes Jason 20 minutes to get dressed,20 minutes to eat and 30 minutes
to walk to school.
Total = 20 + 20 + 30 = 70 min = 1 hour 10 min

Time he should get up = 8:30 – 1 hour 10 minutes = 7:20 a.m.

240

Solution to Question **132**

Peter arrived at work at 15.45 today.
It took him 90 minutes to get there i.e. 1 hour 30 min
Time at which he left home = 15:45 – 1:30 = 14:15 i.e. at
2:15 p.m.

 ## Solution to Question **133**

Gavin went for a walk at 10.25 a.m.
He walked for 2 hours and 37 minutes.
Time at which he returned home = 10:25 + 2:37 =
1:02 p.m.

 ## Solution to Question **134**

It takes a newspaper vendor 2 hour and 46 minutes to deliver his newspapers.
He starts at 7.55 a.m. in the morning.
Time at which he gets home = 7:55 + 2:46 = 10:41 a.m.

 ## Solution to Question **135**

In the last four days, Tim watched six TV programs each day.
Total number of programs watched in 4 days = 4 x 6 = 24
Each program was 35 minutes long
Total duration of 24 programs in minutes = 35 x 24 = 840 minutes
1 hour = 60 min
Time in hours that Tim spent watching TV = 840/60 = 14 hours

 ## Solution to Question **136**

Brad spent 37 minutes on each worksheet he completed.
He completed 17 worksheets.
Total time spent on completing 17 worksheets = 37 x 17 = 629
minutes 1 hour = 60 min
Time in hours that Brad spent in working on worksheets = 629/60
= 10 hours and 29 minutes

241

Solution to Question 137

Kate took 24 minutes to finish the first 12 questions of an exam that has 50 questions in all.
Remaining questions = 50 – 12 = 38
12 Questions ————— 24 min
1 Question = 24/2 = 2 min
38 Questions ————— ? min
At this rate, the time that Kate needs to finish the exam = 38 x 2
= 76 min i.e. 1 hour 16 minutes

 Solution to Question 138

Lisa spent seventy-four minutes preparing for an exam = 1 hour 14 min Jane spent 3 hours and 54 minutes preparing for the exam.
Difference in preparation time of the two = 3:54 – 1:14 = 2 hours 40 minutes
Jane spent 2 hours and 40 minutes longer to prepare for the exam.

 Solution to Question 139

Lily worked out on the treadmill for 55 minutes.
She then went cycling for 1 hour and 25 minutes.
Total time that Lily exercised = 0:55 + 1:25 = 2 hours 20 minutes

 Solution to Question 140

Suzanne was baking a cake for her mother.
It took thirty four minutes to prepare the batter.
Then, the cake baked in the oven for forty-five minutes.
Total time needed for preparation and baking = 34 + 45 = 79 minutes i.e. 1 hour 19 minutes
Suzanne started at 12:43 p.m.
Time by which the cake was done = 12:43 + 1:19 = 2:02 p.m.

 Solution to Question 141

Arnold's flight was scheduled for departure at 9:44 a.m.

242

The flight got delayed by ninety minutes = 1 hour 30 min
The new departure time = 9:44 + 1:30 = 11:14 a.m.

Solution to Question 142

Tina takes 18 minutes to finish each worksheet.
Today she started working at 3:15 p.m. and completed three
worksheets. Time taken to complete 3 sheets = 18 x 3 = 54 minutes
Time at which Tina finished = 3:15 + 0:54 = 4:09 p.m.

Solution to Question 143

Betty is at the airport.
Her flight takes off at 11:17 p.m.
The expected flight time is 4 hours and 45 minutes. Time at
which she would land = 11:17 + 4:45 = 4:02 a.m.

Solution to Question 144

Jack can get to school faster by cycling.
He reaches school by 8:45 a.m. when he cycles and by 10:20 a.m. when he
walks.
Time difference = 10:20 − 8:45 = 1 hour 35 minutes
Jack can be quicker by 1 hour and 35 minutes if he cycles to school.

Solution to Question 145

Jane is going to meet Paul at the movie theatre for the 10:00 p.m. show.
The current time is 5:15 p.m.
Time Jane has until the show = 10:00 − 5:15 = 4 hours 45 minutes

Solution to Question 146

Siya is flying from New York to San Francisco with a stop over at Chicago.

243

The plane will land in Chicago at 1:38 p.m. and will take off for San Francisco at 3:20 p.m.
Duration of time for which Siya will be at Chicago = 3:20 − 1:38 = 1 hour 42 minutes

 ## Solution to Question 147

Amie was invited to Mark's birthday party.
The party started at 4:00 p.m.
Amie reached sixty-five minutes after the party started = 1 hour 5 minutes
Time at which Amie reached = 4:00 + 1:05 = 5:05 p.m.

 ## Solution to Question 148

Jenny hasn't decided if she wants to see a movie at 5:00 p.m. or at 6:15 p.m.
The time currently is 3:30 p.m.
Duration of time for which Jenny would need to wait for the first show = 5:00 − 3:30 = 1 hour 30 minutes
Duration of time for which Jenny would need to wait for the second show = 6:15 − 3: 30 = 2 hours 45 minutes
Difference = 1 hour 15 minutes
Jenny needs to wait for 1 hour and 15 minutes longer for the second show as compared to waiting for the first show.

 ## Solution to Question 149

Mary = 11/01/2001 - 20/11/1942 = 59 years
Benjamin = 04/11/1990 - 13/05/1954 = 36 years
Frank = 13/12/1915 - 01/02/1851 = 64 years
Martha = 11/03/2002 - 28/07/1934 = 68 years
Sandra = 27/06/1954 - 30/06/1871 = 83 years

a) List of family members in order of how long they lived.
Sandra, Martha, Frank, Mary, Benjamin

b) List of family members in order of when they were born. Frank, Sandra, Martha, Mary, Benjamin

c) List of family members in order of when they died.
Frank, Sandra, Benjamin, Mary, Martha

244

Solution to Question 150

a) 1 week = 7 days
I am 1/7 of a week = 1/7 x 7 = 1 day
1 day = 24 hours
I am a unit of time on Earth.
I am 24 hours
b) I am 1/60 of 1/60 of 1/24 of the answer to number one.
Answer = 24 x (1/60) x (1/60) x (1/24) hours
= 1/3600 hours
(one hour has 60 minutes and each minute has 60 seconds)
= 1/3600 x 60 x 60 seconds
= 1 second

Solution to Question 151

William is 7 times as old as his son this year. His son was 4 years old last year.
Age of William's son this year = 4 + 1 = 5 years
Age of William this year = 7 x 5 = 35 years
Difference between William's age and his son's age = 35 -5 = 30 years
When William is 60 years his son would be = 60 – 30 = 30 years

Solution to Question 152

Joshua is 12 years old now. His father is 32 years older than him.
Father's age = 12 + 32 = 44 years
Joshua's age in 5 years time = 12 + 5 = 17 years
Fathers' age in 5 years would be = 44 + 5 = 49 years
Total of Joshua's age and his father's age in 5 years time = 49 + 17 = 66 years

Solution to Question 153

John ran 25 miles.
Henry ran twice as far as John = 25 x 2 = 50 miles
Ben ran 10 miles more than Henry = 50 + 10 = 60 miles
Peter ran eight miles less than Henry = 50 – 8 = 42 miles

245

Solution to Question **154**

Laura weighs 14 kg.
Her first sister is 4 times heavier than Laura = 14 x 4 = 56 kg
Her second sister is half the weight of her first sister = 56/2 = 28 kg
Weight of her 2 sisters put together = 56 + 28 = 84 kg

 Solution to Question **155**

Capacity of 9 bottles = 9 x 726 ml = 6534 ml
3005ml of water overflowed from the jug = 6534 - 3005 = 3529 ml
Capacity of the jug = 3529 ml

 Solution to Question **156**

Tia made a sandwich with some cheese and 2 slices of bread that each weighed 2 grams.
Weight of 2 slice of breadth = 2 x 2 = 4 grams
She weighed the sandwich and found it weighed 9 grams.
Cheese she used = 9 − 4 = 5 grams

 Solution to Question **157**

A 2.5 litre bottle of cola is shared between 5 friends.
Each person gets = 2.5/5 = 0.5 litres = 500 ml

 Solution to Question **158**

Michael drinks a 350 ml bottle of lemonade every day.
Amount of lemonade he will drink in one week = 350 x 7 = 2450 ml
1 litre = 1000 ml
? litre = 2450 ml
= 2450/1000 = 2.45 litres
Amount of lemonade he will drink in one week in litres = 2.45 litres

246

Solution to Question 159

A car uses 3.5 liters of fuel every 2 kilometers it travels.
3.5 liters ——— 2 km
 ? liters ——— 75 km
Amount of fuel it uses to travels 75 kilometers = 75 x 3.5/2 = 131.25 liters

Solution to Question 160

Amie has a jug of lemonade.
 She does not know how much lemonade she has, but she knows she can fill 12 glasses which have a capacity of 280 ml each.
Amount of lemonade she has = 280 x 12 = 3360 ml i.e. 3.36 liters

Solution to Question 161

Jug A holds 1800 ml water.
Jug B holds 3/4 more = 1800 + 1800 x 3/4 = 1800 + 1350 = 3150 ml
Amount of water jug B hold = 3150 ml
Amount of water the two jugs hold = 3150 + 1800 = 4950 ml

Solution to Question 162

Sarah creates a fruit punch.
1 liter = 1000 ml
It contains 1/10 of a liter of apple juice = 1000/10 = 100 ml
2/5 of a liter of orange juice = 2/5 x 1000 = 400 ml
1/8 of a liter of grape juice = 1000/8 = 125 ml
The jug which is the most suitable for Sarah to serve her fruit punch in
= 100 + 400 + 125 = 625 ml i.e. Jug 3

Solution to Question 163

Rick is 98 centimeters tall.
Ted is 31 centimeters taller.

Difference in their heights = 98 − 31 = 67 centimeters
1 m = 100 centimeters
? m —— 67 centimeters
= 67/100
= 0.67 meters
Rick is 67 centimeters or 0.67 meters taller than Ted.

 ## Solution to Question **164**

A piece of wood is 5 meters long.
A carpenter cuts off 45 centimeters.
1 m = 100 centimeters
? m ——— 45 centimeters
= 45/100
= 0.45 meters
Length of the wood after it is cut = 5 − 0.45 = 4.55 meters or 4.55 x 100 = 455 centimeters

 ## Solution to Question **165**

Andy cuts a long pipe into 5 pieces.
Each piece is 45 centimeters.
Total length = 5 x 45 = 225 centimeters
1 m = 100 centimeters
? m —— 225 centimeters
= 225/100
= 2.25 meters
The original length of the pipe in meters = 2.25 meters

 ## Solution to Question **166**

Richard has 40 books.
He needs 1.2 meters of shelving to store them.
1 m = 100 cm
1 cm = 10 mm
Each book's width = 1.2/40 = 0.03 meters i.e. 3 centimeters i.e. 30 mm

248

Solution to Question 167

2.4 meters of ribbon is cut into 4 equal pieces.
Each piece in meters = 2.4/4 = 0.6 meters
1 m = 100 cm
0.6 m ——— ? cm
= 0.6 x 100
= 60 centimeters
Length of each piece in centimeters = 60 centimeters

Solution to Question 168

A carpenter has a plank of wood 3 meters long.
He cuts it into 6 pieces.
Length of each piece of wood in meters = 3/6 = 0.5 meters
1 m = 100 cm
0.5m —— ? cm
= 0.5 x 100
= 50 cm
Length of each piece of wood in centimeters = 50 centimeters

Solution to Question 169

Mike participated in a 3000 meters in a race. He stopped after 1100 m.
Distance that Mike is yet to cover to complete the race =
3000 – 1100 = 1900 meters

Solution to Question 170

Paul dug 9 rows of earth to plant vegetables.
Each row was 4 meters long.
1 m = 100 cm
Length of the ground he dug altogether in meters = 9 x 4 = 36 meters Length
in centimeters = 36 x 100 = 3600 centimeters

249

Solution to Question 171

Julian and her friends wanted to bake a birthday cake for their friend. They needed 3.5 kg of sugar for the recipe.
Sarah had 1.6 kg of sugar and Tim had 1300 g (1.3 kg).
1000 g = 1 kg
Total = 1.6 + 1.3 = 2.9 kg
Amount of sugar they needed to buy more to have enough for the recipe = 3.5 − 2.9 = 0.6 kg

Solution to Question 172

Peter needed 6 kg of sand for his cement mix.
He only had 1.8 kg.
While carrying the sack of sand he dropped it and spilled another 0.9 kg.
Total sand with him now = 1.8 − 0.9 = 0.9 kg
Amount of sand he needs in order to have enough for the cement mix = 6 − 0.9 = 5.1 kg

Solution to Question 173

Sandra wanted to post a parcel to her friend Kathy.
Her parcel contained a 200 grams bar of chocolate, a 350 grams bag of sweets, 2 packets of biscuits that weighed 20 grams each and a cake that weighed 950 grams.
Total weight of all the items = 200 + 350 + 20 + 20 + 950 = 1540 g
Sandra forgot the weight of the parcel box.
The parcel was weighed at the post office and it weighed 1560 grams in all.
Weight of the box = 1560 − 1540 = 20 grams

Solution to Question 174

Bella bought a 3.2kg fruit cake.
Shane came in and cut the cake into half to share with his friends = 3.2/2 = 1.6 kg
Then Paul came in and cut himself a slice.
The cake now weighed 900g.
a) Weight of Paul's piece of cake = 1.6 − 0.9 = 0.7 kg
b) Weight of the cake Shane shared with his friends = 1.6 kg

250

Solution to Question 175

Laura bought a 2.8 kg jar of peanut butter for her Mum.
She took out 700 grams of peanut butter separately from the jar for her friend.
Weight of the jar now = 2.8 − 0.7 = 2.1 kg
Laura bought 3 more small jars of peanut butter to refill the bigger jar for her mother.
Each small jar weighed 225 g.
Total weight of small jars = 225 x 3 = 675 grams = 0.675 kg
The weight of the big jar after it was refilled = 2.1 + 0.675 = 2.775 kg

 ## Solution to Question 176

The dimensions of the picture + the border,
Length = 24 + 4 + 4 = 32 cms
Breadth = 8 + 4 + 4 = 16 cms

Area of the border = Total area −
area of the paper =
32 x 16 − 24 x 8
= 512 − 192

= 320 sq. cm

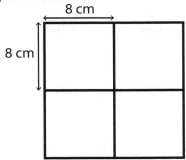 ## Solution to Question 177

4 small squares are used to form a large square.
Each side of a small square is 8cm

8 cm

8 cm

Then the side of large square = 8 + 8 = 16 cm the
perimeter of the large square = 16 x 4 = 64cm

251

Solution to Question 178

Area of the hall = 22 x 17 = 374 sq. m
Cost per square meter = $8.00
Total cost = $374 x 8 = $2992

 Solution to Question 179

Area to be painted = 36 x 36 = 1296 sq. m
Cost of painting the board = 1296 x 4 =$ 5184

 Solution to Question 180

Area of the room = 12 x 25 = 300 square-feet
Area of 1 square foot tile = 1 x 1 = 1 sq. foot
Number of 1- square-foot tiles needed to tile the room = 300/1 = 300 tiles

1 foot = 12 inches
Area of the room in square inches = 300 x 12 x 12 = 43200 inches
Area of a 3 x 3 inch tile = 3 x 3 = 9 square inches
Number of 3 inches tiles required to tile the room =43200/9 = 4800 tiles

 Solution to Question 181

The rectangle and the square have the same perimeter.
Perimeter of the rectangle = 2 x (15 + 25) = 2 x 40 = 80 cm
Perimeter of the square = 80 cm
Length of one side of the square = 80/4 = 20 cm

 Solution to Question 182

1 foot = 12 inches
The square sheet has a side that measures 2 feet = 2 x 12 = 24 inches
Area of the square sheet = 24 x 24 = 576 sq. in
Area of one square card = 2 x 2 = 4 sq. in
The greatest number of cards that can be cut from the sheet = 576/4 = 144

252

Solution to Question 183

Perimeter of the figure is equal to the length of the sections marked on the diagram.

= 4 + 2 + 4 + 2 + 4 + 4 + 4 + 4 + 4 + 2 + 4 + 2 + 4 + 4
= 48 cm

Solution to Question 184

Breadth = 9 cm
Length of rectangle is thrice its breadth = 9 x 3 = 27cm
Area of rectangle = 9 x 27 = 243 sq. cm

Solution to Question 185

Number of full squares = 12
Number of squares more than half full = 6
Number of squares less than half full = 6

Count the number of squares more than half full as 1 and those less than half full as 0

Number of squares = 12 + 6 = 18
Area of each square = 1 x 1 = 1 sq. cm
Area of 'N' = 18 x 1 = 18 sq. cm

253

Solution to Question 186

Area of square = 16 sq. cm
Let the side of square be X
X x X = 4 x 4 = 16
X = 4 cm
Perimeter of the figure =
the length of the 12 sections marked
= 12 x 4
= 48 cm

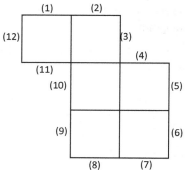

Solution to Question 187

Area of square = 4 sq. cm
Area of the whole figure = 10 x 4 = 40 sq. cm.

Solution to Question 188

Liz has a regular octagon shaped box.
She wants to decorate it with a ribbon all around the edge.
An octagon has 8 sides.
Each side of the box measures 7.54 centimeters.
To calculate the amount of ribbon needed, we need to calculate its perimeter.
Therefore total amount of ribbon she needs altogether = 8 x 7.54 = 60.32
centimeters

Solution to Question 189

Jimmy has been asked to put a fence all around a regular hexagonal garden.
The perimeter of the garden is 46.8 meters.
Hexagon has 6 sides; as it is a regular hexagon all its sides are equal.
Measurement of each side = 46.8/6 = 7.8 centimeters

Solution to Question 190

Cindy had a regular pentagon shaped box.
She decorated it all around the edge with shiny circles. Each circle had a diameter of 2 centimeters.
The perimeter of the box was 200 centimeters.

a) Number of circles Cindy needed altogether = 200/2 = 100
b) A pentagon has 5 sides.

Number of circles she put on each side = 100/5 = 20

 ## Solution to Question 191

a) Farmer Tom has a square field.
Its each side measures 120 meters.
He wants to put a barbed wiring all around it.
Therefore we need to calculate the perimeter of the square.
Amount of wiring he needs = 4 x 120 = 480 meters
b) Perimeter of the rectangular field = 2 x (L + B)
=2 x (21 + 15)
= 2 x 36
= 72 meters

 ## Solution to Question 192

A rectangle has an area of 56 sq.cm. The lengths of its sides are WHOLE numbers. First let us write down the factors of 56 to know the sides of rectangle.
The factors of 56 are 1, 2, 4, 7, 8, 14, 28 and 56.
From the above factors, we can consider the following as the sides of rectangle with area 56 sq. cm.

1, 56
2, 28
4, 14

7, 8
The smallest possible perimeter = 2 x (7 + 8) = 2 x 15 = 30 centimeters
The largest possible perimeter = 2 x (1 + 56) = 2 x 57 = 114 centimeters

255

Solution to Question 193

Judie wants to have a new carpeting for her bedroom.
Her bedroom is a 11 meter by 9 meter rectangle.
Carpeting costs $6 per square meter.
Area of the bedroom = 11 x 9 = 99 sq. m
Total cost of carpeting the entire bedroom = 99 x 6 = $594

 Solution to Question 194

Siya is making a display board for her school.
The display board is a 15 feet by 9 feet rectangle.
She needs to put a ribbon border around the entire display board.
For this we need to calculate the perimeter of the display board.
The length of ribbon that she needs = 2 x (15 + 9 = 2 x 24 = 48 feet
1 foot = 12 inches and 1 inch = 2.54 centimeters
48 feet =
= 48 x 12 x 2.54 centimeters
= 1463.04 centimeters

 Solution to Question 195

Jasmine is making a poster for the school talent show.
The poster is a 0.75 meters by 0.90 meters rectangle. Perimeter
of the poster = 2 x (0.75 + 0.9) = 2 x 1.65 = 3.30 meters
Ribbon costs $2 per meter.
Total cost of adding a ribbon border around the entire display board
= $2 x 3.3
= $6.60

 **Solution to Question 196**

Danny has a rectangular rose garden that measures 800 centimeters by
1200 centimeters.

Area of rose garden = 800 x 1200 = 960000 sq. cm
1 sq. centimeters = 0.0001 sq. m
960000 sq. centimeters ———— ? sq. m
= 960000 x 0.0001
= 96 sq. m
One bag of fertilizer can cover 16 meter square.
1 bag ———— 16 sq. m
? Bags ———— 96 sq. m
Number of bags he would need to cover the entire garden = 96/16 = 6

Solution to Question 197

Ella wants to buy wood to make a frame for her picture.
Her picture is a 12 feet by 10 feet rectangle.
Perimeter of the picture = 2 x (12 + 10) = 2 x 22 = 44 feet
1 foot = 30.48 centimeters
44 ———— ? centimeters
= 44 x 30.48
= 1341.12 centimeters
Total length of the wood strips in centimeters that she would need for her
picture = 1341.12 centimeters

Solution to Question 198

The side wall of my room is 13 feet by 9 feet.
Area of the wall = 13 x 9 = 117 sq. feet.
A can of paint covers 50 square feet of the wall.
It will not be enough to paint the entire wall because the area of the wall
is more than the area that can be covered by one can of paint.

Solution to Question 199

Alex wants to put tiles on his bathroom floor.
His bathroom measures 6 feet by 8 feet.
Area of the bathroom = 6 x 8 = 48 sq feet
Each tile measures 1 square foot.
Number of tiles needed = 48/1 = 48

Solution to Question 200

There are 60 books in a pile.
Each book is 3 millimeters thick.
Height of pile = 60 x 3 = 180 millimeters
10 millimeters = 1 centimeters
180 millimeters ———— ? centimeters
= 180/10
= 18 centimeters
Therefore height of the pile in centimeters = 18 centimeters

258

CPSIA information can be obtained
at www.ICGtesting.com
Printed in the USA
LVHW100949160720
660851LV00004B/58